Leaching Gold and Silver Ores With The Plattner and Kiss Processes

by C.H. Aaron

with an introduction by Kerby Jackson

This work contains material that was originally published in 1881.

This publication is within the Public Domain.

This edition is reprinted for educational purposes and in accordance with all applicable Federal Laws.

Introduction Copyright 2015 by Kerby Jackson

Introduction

It has been over a century since C.H. Aaron released his important publication "Leaching Gold and Silvers: The Plattner and Kiss Processes". First released in 1881, this work has been unavailable to the mining community since those days, with the exception of expensive original collector's copies and poorly produced digital editions.

It has often been said that "*gold is where you find it*", but even beginning prospectors understand that their chances for finding something of value in the earth or in the streams of the Golden West are dramatically increased by going back to those places where gold and other minerals were once mined by our forerunners. Despite this, much of the contemporary information on local mining history that is currently available is mostly a result of mere local folklore and persistent rumors of major strikes, the details and facts of which, have long been distorted. Long gone are the old timers and with them, the days of first hand knowledge of the mines of the area and how they operated. Also long gone are most of their notes, their assay reports, their mine maps and personal scrapbooks, along with most of the surveys and reports that were performed for them by private and government geologists. Even published books such as this one are often retired to the local landfill or backyard burn pile by the descendents of those old timers and disappear at an alarming rate. Despite the fact that we live in the so-called "Information Age" where information is supposedly only the push of a button on a keyboard away, true insight into mining properties remains illusive and hard to come by, even to those of us who seek out this sort of information as if our lives depend upon it. Without this type of information readily available to the average independent miner, there is little hope that our metal mining industry will ever recover.

This important volume and others like it, are being presented in their entirety again, in the hope that the average prospector will no longer stumble through the overgrown hills and the tailing strewn creeks without being well informed enough to have a chance to succeed at his ventures.

Kerby Jackson
Josephine County, Oregon
January 2015

CONTENTS.

CONTENTS—CONTINUED.

LIST OF ILLUSTRATIONS.

PREFACE.

This book is written in the endeavor to supply, in some small degree, a want which is severely felt on this coast, namely, that of plain, practical books on metallurgy. In the standard works on this subject, especially in regard to gold and silver, there is a great lack of those practical details which are so essential to the success of the operator, while some of them are characterized by a display of scientific lore which is very discouraging to those whose preliminary education does not enable them to understand it.

This condition of affairs is probably due, in part, to the circumstance that books are less often written by practical workers in this branch than by scientific gentlemen who obtain such practical details as they do give us, not from their own experience, but by inquiry and observation, more or less extensive. It is also in part owing to the fact that in Europe, where most of the works alluded to were written, there are fewer men who are called on to conduct metallurgical operations without previous apprenticeship to the business.

The processes selected for description, namely, the Plattner for gold, and the Kiss modification of the Patera for silver, are those which seem the best adapted in general to our wants; the first, for the sufficient reason that it is the only available process for the extraction of gold by lixiviation; the second, because it is more convenient, and requires a less extensive plant than the

Augustin process, which depends upon dissolving silver chloride by means of a hot solution of common salt, while the Ziervogel process, depending on the formation of silver sulphate, which is extracted by means of hot water, is only adapted to the treatment of matte.

The Hunt & Douglass silver copper process is a method of lixiviation which has much to commend it, but its mode of operation and field of adaptability are so distinct that it will more properly form the subject of a separate treatise.

In the arrangement of the book, the author endeavors to make the necessary explanations and practical directions as simple and straightforward as possible, while matter which, however interesting or instructive, is not essential, appears toward the end, in a separate division.

While it is deemed necessary to give an outline of the *rationale* of the different operations described, it is also thought desirable to avoid, if possible, the use of terms which might be in conflict with either the old or the new systems of notation. Thus, "sodium sulphate," though implying a departure from the still older binary system, in which the formula for the salt was $Na\,O\,S\,O_3$, and in which it was regarded as a sulphate of oxide of sodium, and called sulphate of soda, is, notwithstanding, compatible with the use of the old atomic weights, and represents a view of molecular constitution which antedates, by several years, the general adoption by chemists of the new weights.

It is with this intention that the term sulphur oxide is used, in preference to either sulphuric and sulphurous acid, or anhydride, in speaking of the products of the

combustion of sulphur, and the formation of metal sulphates; for, while strictly accurate, and sufficiently pre-
cise for the purpose, it is the only intermediate term
which could be used as applicable to either of the systems of notation.

The statement met with in every work on the subject, that "sulphuric acid," by which must, of course,
be understood "sulphuric anhydride," acts directly on
sodium chloride, with evolution of chlorine, seems,
while doubtless true, to require more explanation than
is usually given. It may be that air takes a part in the
reaction, by supplying the oxygen neceessary for the formation of sodium sulphate, but it appears to the writer
that a clue to the true explanation is furnished by Brande's
statement that sulphuric anhydride is decomposed by heat
into sulphurous anhydride and oxygen. The decomposition is probably assisted by the affinities of the
sodium in the gaseous sodium chloride. In this view
the reaction would be

$$Na\ Cl + 2\ S\ O_3 = Na\ S\ O_4 + S\ O_2 + Cl,$$
$$(2\ Na\ Cl + 2\ S\ O_3 = Na_2\ S\ O_4 + S\ O_2 + Cl_2)$$

which can take place within the roasting mass, where
air can have little to do with it, as well as in the atmo-
sphere above the ore, the sulphuric anhydride being
furnished by the decomposing metal sulphates, under
the influence of heat, and the nascent chlorine having
the best opportunity to act on remaining sulphides.

Chlorine which may be evolved, or may rise, above
the surface of the ore will, in presence of sulphurous
anhydride and steam, which latter must be produced
whenever fresh fuel (wood or coal) is introduced,

form hydrochloric acid, with reproduction of sulphuric anhydride.

For valuable assistance in the literary part of the work the author is indebted to his friend, Professor John Calvert, of the California College of Pharmacy, San Francisco. The works which have been consulted comprise those of Regnault, Cooke, Abell and Bloxham, Ure, and Kustel, as well as Lippincott's Encyclopedia of Chemistry.

While the author has not succeeded in entirely satisfying himself, he ventures to hope that the book will be found useful by those who may have occasion for it.

INTRODUCTION.

1. Leaching, or lixiviation, originally meant the extraction of alkaline salts from ashes, by pouring water on them. The resulting liquid was called a leach, lixivium, or lye. In metallurgy, at the present day, leaching means the extraction of metal salts from ores, by means of a watery solvent. The solution so obtained is called the leach or lixivium. Leaching may be done in several ways—by filtration, by decantation, or by flowing.

2. Filtration may be upward or downward; the solvent is passed through the prepared ore and through a filter which retains all solid matter.

3. Decantation is drawing or pouring off the solution obtained by mixing the solvent with the ore, after allowing the solid matters to subside.

4. Flowing is allowing a stream of the solvent to flow continuously into the lower part, and out of the upper part of a vessel containing the ore. The latter is usually kept suspended in the liquid by gentle stirring, and the vessel is so deep that only clear, or nearly clear liquid rises to the outlet.

5. Ores are usually leached for gold and silver by downward filtration.

6. It is necessary that the metals be combined with some substance which renders them soluble in the liquid used. For gold and silver, chlorine is the most suitable

substance for this purpose. In order, then, to extract gold and silver from ores by leaching, these metals are combined with chlorine; the resulting compounds, or metal salts, are dissolved, and the leach is separated from the undissolved matter by filtration. The metals are then separated from the leach by precipitation.

7. Chemical combination differs from mere mixture, or mechanical combination. When two or more substances are merely mixed, each remains the same as it was before, and may be separated from the others without having undergone any alteration of its properties; but when they are chemically combined, they unite to form a substance which usually differs from any one, and from all of the original components—a substance which is the same throughout, or homogeneous, and from which the components can only be recovered by chemical action. Thus gold and chlorine might, under certain conditions, be mixed, and again separated unchanged, by a current of air; in the interval the gold would remain a metal, and the chlorine a gas. If combined, gold chloride would be formed, which is neither a metal nor a gas. It bears no resemblance, in appearance or properties, to either gold or chlorine; nor can either of those substances be obtained from it by any merely mechanical process of separation.

8. Solution also differs from mixture; the difference is best explained by an example. Sand, clay, or gold powder can be mixed with water, but not dissolved in it. They may remain a long time suspended, but will ultimately settle, or may be at once separated from the water by filtering; while sugar, salt, or gold chloride will dissolve in water, will not settle, and will pass with

the water through a filter, having in the act of dissolving become a liquid, which the other substances, not dissolving, did not.

9. Precipitation means throwing down. It is effected by adding to a solution a substance, either solid or liquid, which acts chemically, and causes the dissolved substance, or some of its components, to become insoluble in the liquid; or, so changes the liquid as to render it incapable of dissolving the substance, which is therefore thrown down, or precipitated. An example of the first kind of precipitation is seen when a drop of muriatic acid is added to a solution of silver nitrate. Silver chloride is formed, which, being insoluble in the liquid, separates in the solid state. If this silver chloride be placed in hot brine, it will be dissolved, and will be again thrown down on the addition of sulphuric acid, which so changes the brine as to render it incapable of dissolving silver chloride. This is the second kind of precipitation.

Precipitation may also be caused by a change of temperature. Thus, if the hot brine in which the silver chloride was dissolved were allowed to become cool, the silver chloride would, unless in very small quantity, be precipitated, because cold brine cannot dissolve so much as when hot. Again, as any solvent can only dissolve a certain quantity of a substance, it follows that if a saturated solution be exposed to evaporation, the dissolved substance, not evaporating, must be thrown down in proportion to the diminution in quantity of the solvent. In the latter two cases the effect is generally called crystallization, or deposition, rather than precipitation. A substance which causes precipitation, when added to a solution, is called a precipitant.

10.   Chlorine is a greenish gas, and is usually produced from common salt.   When metals are combined with chlorine, the resulting compounds are metal chlorides, and are distinguished by the names of the respective metals, as gold chloride, silver chloride, and further, by the prefixes sub, or di, proto, bi, ter, tetra, penta, as copper chloride, copper protochloride, mercury protochloride, and bichloride, etc., representing different proportions of chlorine combined with the respective metals.   The terms sesqui and per, are also used to designate certain ratios of combination.

11.   Gold terchloride, for lixiviation, usually called simply gold chloride, is made by exposing the pulverized ore, containing the metal in small particles, to the action of chlorine and moisture.   It is extracted by leaching with cold water, in which it dissolves readily, and the gold is precipitated in the metallic state, as a brown powder, by a solution of iron sulphate, known in commerce as copperas, or green vitriol, which takes to itself the chlorine, and leaves the gold insoluble.   The metal is collected, washed, dried, melted, and cast as a bar or ingot.

12.   If the rock, or ore containing the gold, is free from opposing substances, it may be chlorinated without being roasted; but in general, ores which are treated by lixiviation contain the gold so combined, or mixed with other substances, that a preliminary roasting is necessary.

13.   Silver chloride is made, for lixiviation, by means of heat, in a roasting furnace, is extracted by leaching the ore with a solution of calcium hyposulphite, being

insoluble in simple water, and the silver is precipitated as silver sulphide, by a solution of calcium polysulphide. The precipitated sulphide, in the form of a black mud, is collected, washed, dried, roasted, and melted with an addition of scrap-iron, which takes the sulphur remaining after the roasting, and sets the silver free.

14. Silver sometimes exists naturally combined with chlorine, in ore, and is then soluble without roasting; but in general it is combined with antimony, sulphur or arsenic, or with base metal sulphides, oxides, etc., and then roasting is necessary.

ROASTING.

15.   As a metal chloride is a compound of a metal with chlorine, so a metal oxide is a compound of a metal with oxygen, and a metal sulphide, or sulphuret, is a metal combined with sulphur, while a metal sulphate is a metal combined with both oxygen and sulphur—that is, a metal oxide with a sulphur oxide; the latter being the same which, when combined with a certain proportion of water, is called sulphuric acid, or oil of vitriol. The metal oxides, sulphides, and sulphates are distinguished in the same way as the chlorides.

16.   The purpose and effect of roasting ore for lixiviation is, as to gold, to burn all base metals, sulphur, and other substances, such as arsenic, antimony, and tellurium, and either expel them by volatilization, or leave them in such condition as to be harmless in the chlorination of the gold, and, as to silver, to change its condition in the ore, from various insoluble compounds, into soluble silver chloride.

17.   In order that roasting may be effective, the ore must first be crushed to powder. The most suitable degree of comminution must be found by trial for each particular ore. The more coarsely it is crushed, consistently with good roasting, the more easily is it leached. A powder which will pass through wire gauze of 40 meshes to the running inch is fine enough, and, in some cases, a sieve of 20 meshes to the running inch may be used with advantage. For crushing there is,

as yet, nothing better in the market than the stamp battery.

18. *Oxidizing Roast.*—The crushed ore is exposed to heat, with abundant access of air. The metal sulphides take fire and burn, both metal and sulphur being oxidized by combining with oxygen from the air. A part of the oxidized sulphur flies off with the well known sulphurous smell. Another part combines with a portion of the oxidized metal, forming metal sulphate. The rest of the metal remains as oxide, except silver, which, if not converted into sulphate, becomes metallic. In this way iron, copper, zinc, and lead sulphides are changed, partly into the respective sulphates, and partly into oxides. Nearly the whole of the silver is converted into sulphate, or reduced to the metallic state; gold remains unchanged. Antimony and arsenic are oxidized, and partly fly off, while a part remains, to combine with metal oxides, forming antimonates and arsenates, much in the way in which sulphur makes sulphates. An oxidizing roast is a roast so conducted that the gold is metallic; the silver is either metallic or in the form of sulphate, and the base metals are converted into sulphates or oxides.

19. *Dead Roast.*—Under an increase of heat, some of the metal sulphates which were formed during the oxidation are decomposed; sulphur oxide flies off, and metal oxide remains, although some of the metal oxides also volatilize to some extent. The order in which some of the principal sulphates are decomposed is, iron, copper, silver; the last requiring a very high heat. Lead sulphate is not decomposed, nor is it usual to push the heat so far as to decompose silver sulphate. A dead

roast is an oxidizing roast, carried forward to decomposition of iron and copper sulphates.

20. *Chloridizing Roast.*—At the commencement this is the same as an oxidizing roast; but salt is mixed with the ore, either at the time of charging the furnace, or at a certain stage of the operation. The quantity of salt used depends on circumstances, and varies from one to twenty per cent. of the weight of the ore.

21. Salt is a compound of the metal sodium with chlorine, and is the cheapest source of chlorine. The chlorine of the salt is transferred, under the action of heat, from the sodium to the other metals, by a variety of agencies, chiefly by means of sulphur and oxygen, for which the sodium has a greater affinity, whence it happens that, when another metal sulphate is heated in contact with sodium chloride, an exchange takes place, the sodium takes the sulphur and oxygen, and forms sodium sulphate; the other metal takes the chlorine, and forms a chloride. In this way, iron, copper, zinc, lead, and silver sulphates form chlorides, while a corresponding proportion of salt forms sodium sulphate. For this reason, sulphur is necessary in a chloridizing roast, since without it sulphates cannot exist.

22. The higher sulphur oxide, formerly called dry sulphuric acid, but now known by the name of sulphuric anhydride, also plays a prominent part in the decomposition of the salt and the evolution of chlorine. The latter acts upon any remaining sulphides, and to some extent on oxides, converting them into chlorides. It also acts on metallic silver.

23. Among other agencies involved in the formation of metal chlorides, is that of steam, from burning fuel.

Steam, in contact with salt and quartz, at a red heat, produces hydrochloric acid, which assists in the work. It also decomposes some of the base chlorides, especially the volatile ones, thus giving the silver the benefit of their chlorine. For this reason, steam is sometimes, in the case of rich ores, admitted to the roasting chamber by means of a perforated pipe laid in the firewall, which is made hollow, with openings on the side next to the ore. This, however, causes an increased consumption of fuel.

24. As the heat still increases, the base metal chlorides are decomposed, just as the sulphates are decomposed in the dead roast. They give off the whole, or a part of their chlorine, remaining, or volatilizing, as chlorides of a lower degree, or taking oxygen from the air and becoming oxychlorides, or oxides. Silver chloride is not decomposed by heat.

25. Iron perchloride, formed quite early in the roasting, is volatile, and some of it flies away, while another part, giving off chlorine, is reduced to protochloride, and this again, losing the remaining chlorine, takes oxygen from the air, and forms iron peroxide, which remains in the ore.

26. Copper protochloride gives off half its chlorine, becoming dichloride, which volatilizes to some extent, imparting a deep blue color to the flames.

27. Lead sulphate requires a rather high heat to convert it into chloride, so that it remains in part, sometimes wholly, unchanged. The chloride gives off some chlorine and takes oxygen, becoming oxychloride.

28. Zinc volatilizes partly, in some form, and produces in the flues hard concretions of oxide, or carbon-

ate, which must be removed from time to time. The oxide is quite stable.

29. Antimony and arsenic form volatile chlorides as well as oxides, which, to a great extent, go up the chimney.

30. Gold forms, at a very low heat, a peculiar chloride which decomposes at a higher temperature, and then remains metallic, except as a portion of it may be again chloridized, in the same form, by chlorine from decomposing base chlorides, during the cooling of the ore, after being withdrawn from the furnace. The gold chloride formed in roasting contains less chlorine than that formed by the cold gas, and is not soluble in water, but dissolves in the solvent used for the silver leaching. The formation of this compound cannot be relied on as a process for extracting gold.

31. The presence of lead is disadvantageous for the roasting, because its compounds melt too easily, and because the oxide and chloride volatilize, to the injury of the workman's health; and for the leaching, because the sulphate and chloride are soluble in the silver leach, and the sulphate cannot be removed from the ore by washing with water; as, however, the chloride is soluble in hot water, it is preferable to the sulphate, as it can be removed from the ore before the silver extraction begins.

32. Silver chloride is not very volatile by itself, but in some cases becomes so, apparently from the influence of base metal chlorides, notably that of iron perchloride, when too much heat is used early in the roasting. Antimony and zinc also tend to cause volatilization of silver; so as it does not require a very high heat to

form the silver chloride, the roasting is conducted, not only with a very low heat at the beginning, but without an extremely high temperature at any time, and even a moderately high degree is maintained for a short time only towards the end. A chloridizing roast is a roast which, beginning with oxidation, ends by leaving, as nearly as possible, all the silver in the form of chloride in the ore.

APPARATUS.

33. ROASTING FURNACES—The essential condition of roasting pulverized ore is, that every particle shall be exposed to the action of heated air until certain chemical changes are effected, after which heat alone will complete the operation. This condition is fulfilled in the reverberatory furnace, in which a layer of ore of a certain thickness is acted on by a current of heated air from a fire. The surface of the layer is renewed from time to time by stirring the mass by means of implements operated by hand. As stirring by hand power is laborious and expensive, various means have been devised to dispense with it, and to substitute the automatic action of machinery, among the best of which are the furnaces known as the Stetefeldt, Bruckner, Brunton, White, Howell White, Pacific, and O'Hara, each of which has certain advantages over the other, so that a decided preference, under all circumstances, can be given to neither, and as the reverberatory is the original of all, as any kind of ore that is fit to be roasted can be roasted in it, and as the operator who knows how to use it, can easily adapt himself to the others, a complete description of its construction and operation will be given, but, as a work of this character would be imperfect without some account of the mechanical furnaces also, they will be briefly described in separate articles under "Addenda."

34. *A Reverberatory Furnace* is simply an oven. There is a fire-place at one end and a flue at the other, and the ore to be roasted is laid on a horizontal hearth between them. In the side walls are openings, which admit air, and allow the workman to turn the ore over, from time to time, with a long hoe, or rake. These openings can be closed, when necessary, with iron doors. The fuel must be such as will produce a flame; hence, wood, or flaming coal is suitable.

35. Reverberatory furnaces are of several kinds; the single, which has but one roasting chamber, the double, in which a second roasting chamber is constructed directly over the first, and is worked with waste heat; and the long furnace, which has two or more chambers, either in one horizontal plane, or, which is better where the ground is suitable, raised more or less one above another, in step form, but not superposed.

The single hearth is wasteful of fuel; the double hearth is inconvenient, so I will describe a long furnace with two hearths, represented by the horizontal section, Plate 1, and vertical section, Plate 2. If more than two hearths are required, which with concentrated sulphurets may be the case, a third or fourth can be added, but it will, in general, be necessary to add also an auxiliary fire-place to aid in heating them. The capacity of this furnace is three tons of average silver ore per twenty-four hours, or from one to one and a half tons of concentrated sulphurets in the same length of time.

36. The masonry may be built entirely of common bricks; with adobes in case of necessity; but it is better to make the inside of the fire-place, and the arch over it, of firebrick. If convenient the outside walls can be

of any kind of stone, but, unless a good firestone can be procured for the inside work, that must be of brick. Lime mortar may be used with advantage for the outside work, but all parts which are exposed to much heat must be laid in clay. The brickwork consists of headers and stretchers alternately, appearances being sacrificed for the sake of strength. The masonry below the hearth is not solid, but the space inclosed by the walls is filled with sand, or with earth well tamped, and on this the hearth, or sole, is laid, of the hardest bricks on their narrow sides, without mortar, but afterwards grouted.

37. The walls are supported against the thrust of the arch as follows. At the points shown in the diagram are vertical backstays or "buckstays," of wrought iron, two inches wide by one and a quarter inch thick, and long enough to reach from below the hearth to just above the arched roof. Through holes punched in the backstays, near the ends, project the nutted ends of tie rods of $\frac{3}{4}$ inch round iron. The lower tie rods, passing through the body of the furnace, below the hearth, and uniting the lower ends of opposite backstays, are put in before the hearth is laid. The upper ones extend across the furnace above the walls, uniting the upper ends of the backstays, and are not put in place till after the walls and arch are laid. Usually the furnace is stayed in the direction of its length, in the same manner, except that the lower longitudinal tie rods do not pass entirely through, but are comparatively short, being securely anchored in the body of the furnace, under the hearth. If the ends of the furnace are made masssive, these ties may be dispensed with, especially if the ends

are supported by wooden posts well braced. The cross ties, which are indispensable in some form, may be of wood, uniting the upper ends of strong posts, the lower ends of which are sunk in the ground, close to the side of the furnace, and which replace the iron backstays, the objection being that the posts cannot be set very near the doors, both on account of being in the way, and because of the heat; but near the doors is precisely where they are most needed. Cast iron ribbed backstays are also used, but are liable to break.

38. When all is ready for raising the walls above the hearth, the cast iron door frames are set up, in the middle of the thickness of the wall, which should not be less than 16 inches thick, and a stirring-hoe is passed through each of them in succession, and its range of operation ascertained by trial, and marked on the hearth. The walls are then built up along the marks, cutting off the corners of the otherwise rectangular hearth, and giving it the form seen in the diagram. This, though entailing some loss of space within the furnace, is absolutely necessary, in order that every part of the hearth may be accessible with the hoe.

39. The door frames are two feet long by eight inches high in the clear, are set two inches above the hearth, and are built into the wall as solidly as possible. An exception is made as to the back door on the first hearth, when it is to be used for discharging the roasted ore. It must then be set flush with the hearth, and is two inches higher in the clear, to allow of using a large hoe for discharging, but when the discharge is effected through a trap in the hearth, this door frame is the same as the others. The masonry is arched over the frame,

and as the wall at this point is liable to be rather thin, it is a good plan to arch the skewback a little, in a horizontal plane inwards, by which the thrust of the main arch against this weak part is lessened, and to some extent transferred to the points at each side, where it is received by the backstays and tie-rods; notwithstanding which, it is also well to place above each door frame, outside of the wall, a flat bar of iron two inches wide and half an inch thick, the ends of which are tucked between the adjoining backstays and the wall.

40. Doors are sometimes made of cast iron, and pivoted to the frames, but sheet iron doors detached from the frames answer very well, and are furnished with long handles, made of iron rods riveted to them at right angles; when in place they rest against the rabbet formed by the junction of the frames with the masonry, the projecting handles being supported by the roller in front. Doors sliding vertically on the outside of the wall are very convenient, being counterpoised by a weight attached to one end of a rope which passes over a pulley, the other end being attached to the door by a short chain or a link. A slot in the lower part of the door allows it to be closed while the hoe rests on the roller, the handle of the hoe being supported horizontally by a hook depending from a beam above.

41. When the walls are high enough above the hearth, which is 17 inches at the points opposite the highest part of the arch, and not less than eight inches at the lowest points, the skewback is set up with bricks on end, the lower end being cut to the proper angle, and then the walls are continued up, level with the top of

the arch. All openings in the walls are then temporarily stopped, and the enclosed space is filled with moist sand, up to the skewback along the sides and ends, and higher in the middle, and this is carefully shaped to the required form of the furnace top. On the sand, as a support, the arched roof is laid dry, the bricks on end, one course thick, working from the skewback all along on both sides, and keying in the centre line. All openings in the arch, such as flue or feed holes, are circular, formed with a course of "rollers." When the arch has been well keyed, and the openings filled with sand, it is wetted by pouring water on it, and then grouted with a mixture of clay and sand, thinned with water so as to run into and fill all the interstices between the bricks. The best way to lay the bricks in the arch is the style called "herring bone," but it requires well porportioned bricks, of which the width is just twice the thickness, otherwise straight courses are to be preferred.

As soon as this is done, the backstays and the upper tie rods are placed in position, and all the nuts tightly screwed to support the walls, yet not so much so as to move them. The doors, which were temporarily stopped with loose bricks, or pieces of board to retain the sand, are opened, and the sand, on which the arch was laid, is drawn out, to allow the arch to settle as it dries. In a single hearth furnace, the arch has very little spring transversely, because it abuts the other way against solid end walls, but in a long furnace a transverse spring is more necessary, especially near the junction of the two hearths, where the roof descends and rises again, so as in fac to be an inverted arch, as to the longitudinal section, as may be seen in Plate 2, but, as this is almost the narrowest part of the furnace, there is no danger of its falling, if it has a good spring crosswise.

42. In building the fireplace many masons, very improperly, let the bearing bars for the grates rest on an offset in the brickwork, so that though space be left for the lengthening of the grates by expansion when heated, yet that space is soon filled with ashes and other debris, so that the grates are forced to bend or "buckle.." When they become cool again they shorten but do not straighten, the spaces at the end, left by their shrinkage, are refilled, and, when again expanded by heat, the grates buckle still more and are soon ruined.

The end walls of the ash-pit should be perpendicular, without offsets; the bearing bars should be well clear of them, and the ends of the grates should also clear the walls by at least half an inch. The space thus left for expansion is then bottomless, and remains always open, and the grates remain straight.

43. The flue-holes are connected by flues in any convenient way, with a stack which is two feet square inside, and 20 to 30 feet high. In some part of the flue or stack is a damper, similar to that of a stovepipe, operated by the roaster through the agency of a cord or wire.

44. It is desirable to utilize as much as possible the waste heat from the furnace, and it is a good plan to carry the flue in front of the stamp battery, if dry crushing is done, and there enlarge it to form a dust chamber. The top of the chamber is of sheet or boiler iron, which forms a drier on which to dry the ore for crushing. Even

where only concentrated sulphides are treated, so that no crushing is required, a drier is convenient, and may very properly be the top of the dust chamber. The walls are built two bricks high above the iron plate, and topped with two-inch plank, held down by anchor bolts built into the walls. If, however, there is a battery, the side of the drier next to it is without a wall above the plate, which, if thin, is secured by an anchor bolted iron strap along the edge.

45. If silver ore is treated, a drier is necessary for the precipitate. It may be made as described above, to be heated by the waste heat from the roasting ore, by steam, or by a special fire. A small roasting furnace is also requisite for roasting the dried precipitate. It is built similarly to the large furnace, except that it has but a single hearth and one working door. A hearth containing 36 to 40 square feet of surface, will suffice for the roasting of from one to two thousand ounces of silver, in the form of precipitate, in each twenty-four hours.

46. It will be observed, by those accustomed to furnaces, that there are two small innovations in the plan given. Firstly, the ash-pit is open entirely across the furnace. This gives the operator the choice of leaving it so, or of closing either end, which is sometimes an advantage on account of the draft. It is generally preferable, though contrary to custom, to have the opening on the rear side of the furnace, that is the side opposite to the fire door, because the cold air, entering under the hotter end of the fireplace, tends to equalize the heat. Especially is this the case when the workman pushes the half burned wood back, when introducing fresh fuel,

instead of drawing it forward as he should do. Secondly, the doors on the first hearth are not placed in the middle of its length, as is usual, but a foot nearer to the fireplace. Here again the object is to equalize the heat, by causing the cold air which enters by the door, to pass over the hotter portion of the ore near the fire wall. Another advantage is that the hearth is made wider near the fire wall, and narrower at the other end, thus concentrating the heat toward the part which is farthest from the fire.

47. When a long furnace, which may have any required number of hearths, is built on a hillside, it is a good plan to make each successive hearth two feet, or even more, higher than the preceding one, the one next to the fire-place being the lowest. By this plan the cost of grading is lessened, and the dropping of the ore from one hearth to the other assists greatly in the oxidation. This is called a "step furnace."

48. Furnaces are often built with an arched chamber under the first hearth, as shown by the dotted line in Plate 2. This chamber is closed on the working side of the furnace and open on the other, and the ore is discharged into it through an opening about a foot in diameter, in the hearth near the working door. The opening is closed by an iron plate, which rests on a rabbet a couple of inches below the level of the hearth, and the depression thus formed is filled with roasted ore. A small flue from the chamber, leading through the wall to the interior of the furnace, removes the fumes rising from the hot ore, which is not drawn out of the chamber to the cooling floor for some time.

It is a good arrangement. The floor on each side of a furnace is paved with bricks to a width of 14 feet.

49. *Furnace Tools.*—The tools required for such a furnace are: three or four hoes, each fourteen feet long; a couple of smaller ones about six feet; two spades not shorter than the large hoes; a poker, which, if wood is the fuel, should be made like a boat-hook. The shanks of the hoes and spades are of three-quarters or one inch gaspipe, except some three feet next the head, which is of solid round iron. The heads are of one-quarter or one-half inch boiler iron. For very heavy ores, containing arsenic and antimony, cast-iron hoe heads are used, also rakes. These tools are represented in Plate 2.

50. *Crosby's Furnace.*—At Nevada City, Professor Crosby uses, in connection with a reverberatory furnace, an inclined, rotating, unlined iron cylinder, which receives the concentrated sulphides at the higher end, and delivers them at the lower end, already to a great extent oxidized, into a chamber where they are exposed to the heat from the reverberatory hearth, to which they are removed periodically, in batches, and finished under hand stirring with hoes. A fire is used under the cylinder, to commence the burning of the sulphides, after which they continue burning without such aid, combustion being supported by the air which enters through the open upper end of the inclined cylinder with the sulphides, the fumes being carried off at the other end by the draft of the furnace, with which the cylinder is in communication.

The sulphides, thus burning spontaneously, receiving an abundance of fresh air, being constantly moved by the rotation of the cylinder, and afterwards lying in the furnace, exposed to a higher heat while accumulating, require, when transferred to the finishing hearth, but a short time to complete the roasting. Thus three tons of material which requires 24 hours in an ordinary long furnace, can be turned out daily by the labor of only two men. Some power is consumed in turning the cylinder, which however, when power is not required for other purposes, might be had by applying the waste heat of the furnace to a small steam or hot air engine.

51. *Leaching Vat.*—The leaching is done in wooden tubs, which are coated inside with a mixture of coal tar and asphalt melted together, and applied whilst hot. As the chlorination of gold is also effected in these tubs, they are provided with covers when that metal is present in the ore. If there is no gold, covers are not needed, nor is the coating of the tubs with tar so necessary, being in fact inadmissible if hot water is to be used to wash the ore.

The side of a leaching vat is either vertical, or flaired so that the top of the tub is wider than the bottom. The reverse form is not suitable, because the ore, in settling, draws away from the sides, and leaves a space, or, at least, a greater looseness, through which the chlorine can pass upward, or liquids downward, without passing through the mass. The vat represented in Plate 3, is suitable for the treatment of ore containing both gold and silver. Its capacity is two and three-quarter tons of roasted ore.

The vats have filters near the bottom, and rubber pipes, connecting under the filters, for the solution to flow through to the precipitating tubs. For the admission of chlorine a leaden nipple is inserted in the side of the vat immediately below the filter.

52.   In some works the vats are suspended on iron gudgeons, attached to their sides, in order that they may be emptied quickly by dumping.   It is a convenient arrangement if completed by having a stream of water in a sluice below, or a tramway, for the removal of tailings.   In others the leaching vats themselves are mounted on wheels, and can be trundled to the dumping place.

53.   The filter consists of a false bottom of inch boards, through which half inch holes are pierced at intervals of about four inches.   The boards are laid loosely, with open spaces a quarter of an inch wide between and around them.   The false bottom rests on strips of wood, by which it is raised from half an inch to an inch above the true bottom.   As the vat is slightly inclined toward the discharge side, to insure complete draining, the strips are made thicker at one end than at the other, so that the false bottom is horizontal.   They do not touch the sides of the vat, but leave a space for the flow of solution and diffusion of chlorine.   On the false bottom is a layer of pebbles as small as may be without falling into, or through, the holes.   Over the pebbles is a sheet of burlap, or a layer of old grain sacks, which are cheap and good enough, as they are soon destroyed by chlorine, on which account some operators prefer a layer of fine gravel covered with sand.

54.   *Suction Pipe.*—In cases of difficult leaching, filtration may be facilitated by means of a suction pipe. This is simply the discharge pipe of the leaching vat,

made of stiff hose, or of wood, instead of soft rubber as in other cases, and extended to a vertical depth of from six to twenty-five feet. The hose, near its lower end, is coiled once around, as in the accompanying diagram, and secured by a piece of wire; or, a re-curved piece of lead pipe may be inserted in the end of the discharge pipe, or again, the end may be immersed in a cup of water, although this plan is less convenient than the others. The object, in either case, is to prevent the entrance of air.

55. *Vent Pipe.*—Though not the general practice, it is well to have a vent pipe to prevent disturbance of the filter by the air or gas beneath it, when displaced by the entrance of water, especially if the discharge

[SUCTION AND VENT PIPES.]

pipe is hung up, closed as in the case of using suction, or occupied by the introduction of water below the filter, as in commencing the washing of silver ore. There are several methods of arranging a vent pipe. Perhaps the simplest is the following, represented by a dotted line in the preceding diagram.

When the ore vat is disconnected from the chlorine pipe a short piece of rubber tube is connected at one end with the nipple through which the chlorine was introduced, and at the other with the upper part of the vat, by being inserted tightly in a hole bored through the side, just below the cover. The short tube may be left permanently on the lead nipple, and connected with the chlorine pipe when required, by means of a short piece of lead pipe, which afterwards serves also for connecting the tube with the hole in the vat side.

Whether water be introduced from above or below the ore, the air or gas beneath the filter will pass through the tube, and return to the vat above the ore. If the chlorine nipple is to be used thus as a vent, it should be set in as high as possible, consistently with the delivery of the chlorine below the filter. A special vent pipe may be made by boring a one-quarter inch hole lengthwise through a strip of wood, and attaching the strip to the inside of the vat by means of wooden pins. It should terminate at one end immediately below the vat cover, and at the other, beneath the burlap on the gravel. The upper end must be plugged, and a transverse hole bored to connect with the vertical passage. This aperture must be plugged during the chlorination, and opened when water is admitted.

56. *Sieve.*—This is made of stout brass wire, and has from four to eight meshes to the running inch. It is of an oblong form, about two feet wide by three feet long, and is framed with wood. The sides are six inches high, and are prolonged so as to form handles at one end, like those of a wheelbarrow, and at the other, points of attachment for suspension ropes from a support above. It is operated by hand, being swung back and forth, directly over the vat which is in course of being charged, into which the ore which passes through

falls, in a loose and open condition, while the lumps are retained.

57. *Precipitating Vat.*—Figure 1, Plate 4—This is also a wooden tub, and, if for gold, is coated inside as the leaching vat. A smooth bottom is made in it, either by means of a layer of tar and asphalt, melted together in such proportions as to harden on cooling, or by a bed of Portland cement, which can be shaped as desired, to facilitate the removal of the precipitate. A precipitating vat is made wider at the bottom than at the top, in order that the precipitate may not settle upon the staves.

58. For drawing off the waste liquor after precipitation, the most convenient, because self-acting arrangement is a piece of two inch hose, drawn water-tight through the side of the tub near the bottom, and long enough on the outside to lead to a filter, or a settling tank, on a lower level, and on the inside to reach the top of the tub, where it is secured, when not in use, by a wooden clamp. On this end is a wooden float which causes the hose, when in use, to draw always from the surface; also serving to prevent its drawing too near to the bottom so as to cause a loss of gold. The other opening, furnished with a large wooden faucet, is only used when collecting the precipitate.

The silver precipitating tubs must be larger, or more numerous, than those for gold, because the volume of silver solution is much the greater. The size represented in Plate 4, is sufficient for the gold lixivium from one vat, such as is shown in Plate 3, but both for gold and for silver, extra vats are provided for very weak solutions, drainings, etc. The smaller vat,

Figure 2, Plate 4, is to contain a solution of iron sulphate. It is provided with a filter, and a discharge pipe which reaches to the gold precipitating tub, and which, when not in use, is turned up and fastened as in the figure. This vat is elevated so that the precipitant may be used with convenience. For silver a similar vat contains the calcium polysulphide. It does not require a filter, but the pipe is inserted a little. higher, to allow room for sediment.

59. *Troughs* for conveying the lixivium from the leaching vats to the precipitating tubs, are not built, but hollowed out of timber. They are about five inches wide and four deep, and are tarred.

60. *Well.*—As the silver leaching solution, or "hypo," is not thrown away, but is used again continually, it is usual to provide a well on a lower level than that of the silver precipitating tubs. A tub or square vat sunk in the ground answers the purpose.

61. *Pumps.*—For elevating the leaching liquid from the well a wooden pump is used. The small quantity of metal used in its construction, for fastening the valves, etc., is not injurious. In case, however, a pump is used to raise weak lixivium from the gold leaching, it must be constructed entirely without metal, except lead, in those parts which come in contact with the lixivium. Rubber ball valves are best for this pump. A well may be dispensed with if desired, by having the discharge pipes of the silver precipitating tubs connected directly with the pump,

62. *Filters.*—To receive the precipitate from the silver tub there is a filter vat, either round or square, in which it is drained and washed. The filter is made

like those in the leaching vats, except that, instead of burlap, cotton drilling is used. Filters for the precipitated gold are simple pointed bags made of drilling, or sheets of drilling secured over the rims of water buckets which have plug holes near the bottoms, with hollow plugs through which the filtrate flows to a receptacle beneath. A large tank, in which is a filter made of sand or saw dust, is usually placed outside the works. It is used for filtering the liquid drawn from the gold tub after the precipitation of the gold.

63. *Press.*—To facilitate the drying of the silver precipitate it is pressed into cakes. The press is similar to a cheese press. It is sometimes operated by means of a screw, but a weighted lever is better, because self-acting. A press may be dispensed with if the drying facilities are ample.

64. Pipes and faucets must be provided for the conveyance of water and hypo to the points where they are required. Iron pipes with brass faucets may be used for the hypo. A few paper buckets will be useful.

65. *Chlorine Generator.*—Figure 1, Plate 5, is a section of this apparatus. It is made almost entirely of lead, sometimes heavy sheet, but better cast lead, and is arranged so as to be heated from below, either by a special fire, or preferably, when convenient, by steam. If heated by a fire, it stands on a sand bath, which forms the top of the fire place. If by steam, the generator itself forms the top of an iron steam chest, being supported on strips of wood, which permit the steam to circulate beneath it.

66. The joints of the lead work must not be soldered, but "burned;" that is, joined with melted lead, by means

3

of a gas blow-pipe.   Very coarse solder will, however, answer for repairing it.   Portland cement also answers, in case of necessity, for stopping a leak.

67.   The apparatus consists of a leaden tub *a*, in which the materials are put, and which is surrounded by a water chamber *b*, also of lead, the outer wall of which rises as high as the top of the cover, *c*, and in which the curtain of the cover rests, forming what is called a water joint.   The water should be six inches deep, or more.   The pipe, *d*, is for the removal of spent material.   In the cover is a central water-joint opening, through which passes the stem of a stirrer, *e*, made of iron, covered with lead, and to which is attached a small leaden bell, or curtain, *f*, which, reaching near to the bottom of the water chamber, closes the opening, while allowing the stirrer to be turned round by means of the cross bar, *g*.   Another opening, *h*, which is closed either by a water-joint cover, or, as shown in the figure, by a wooden plug wrapped in greased cloth, serves for introducing material without lifting the main cover.   For the introduction of acid, there is a leaden pipe, *i*, which is coiled once round, as shown in the figure, in order that the lower part of the coil, remaining always full of acid, may act as a valve to prevent the escape of the chlorine.   The pipe, *k*, is for the exit of the chlorine.

The subjoined diagram represents the cover in a different position, showing the lugs and the chain sling, by which it is raised; also a hook sustaining the stirrer, which might otherwise injure the cover.   A lever or tackle is used for the lift-

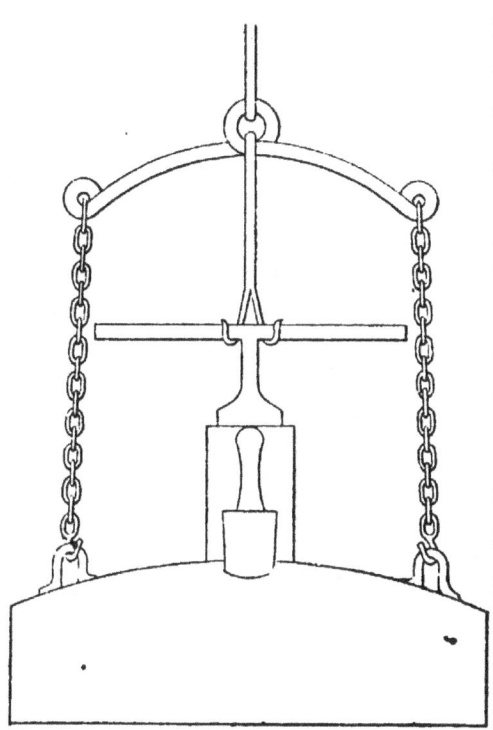

ing. For further remarks on generators, see "Addenda," near the end of this volume..

68. *Wash Bottle.*— Figure 2, Plate 5. This is a bell glass, made from a large bottle, the bottom of which is cut off. It is fixed air-tight, as shown, in the head of a keg, or the tightly fitting cover of a small tub. A piece of pure rubber tube, *l*, which is connected with the leaden pipe, *k*, of the generator, passes through the neck of the bottle, and is packed so as to be air-tight, with oiled rags and putty, and terminates two or three inches below the surface of the water with which the tub or keg is filled. The chlorine, rising through the water into the bell, escapes through the leaden pipe, *m*, and is conveyed to the chlorinating vat by the rubber tube, *n*. The pipe, *m*, is passed, water-tight, through a short piece of oiled rubber tube inserted in a hole bored in the side of the keg, and its upper end, reaching nearly to the top of the bell, and well above the water, is bent downward to prevent the entrance of water, which might occur in consequence of the bubbling up of the gas, and which would cause an obstruction in the rubber tube.

If two vats are chlorinated at once, the rubber tube,

n, is connected with a forked piece of lead pipe, each branch of which is connected with the lead nipple of a vat by a piece of rubber tube. The connections are quickly made by drawing the tightly fitting oiled rubber tubes an inch on to the pipes.

69. All rubber tubes used for chlorine should be oiled on the inside. It is a valuable fact, for the knowledge of which I am indebted to E. N. Riotte, that oil is a great protection against the action of chlorine, which forms with it a white, waxy substance. For this reason, oil or grease is an excellent material for stopping small leaks in the apparatus.

70. The wash bottle serves the double purpose of washing the gas to remove any acid which may pass over with it, and of showing whether or not the generator is operating properly, by the quantity of chlorine which is passing. The chlorine should rise through the water in a steady stream. If it does not, the reason is either that it is not produced fast enough, or the orifice of the pipe is too large, which may be remedied by flattening the pipe more or less. Other forms of wash bottle will be found described under "Addenda."

SPECIAL DIRECTIONS FOR WORKING.

71. *A.—Concentrated Pyrites containing gold, but no silver.*—This material is produced in the gold quartz mills, by crushing the rock in a wet stamp battery, in which the free gold is amalgamated with quicksilver, and passing the tailings over concentrating machines, by which the sulphides are almost entirely freed from rock and earth. It generally consists mainly of iron bisulphide, called pyrites, but often contains copper pyrites, sometimes lead and antimony sulphides, etc., and in other cases consists chiefly of arsenical iron pyrites. Tellurium is also met with.

72. Concentrations should be treated as soon as possible after production, or kept in tanks, under water, for, if allowed to lie long exposed to the weather, they oxidize spontaneously to a certain extent, and form hard lumps which must be repulverized for roasting, for which purpose a self-feeding and discharging barrel pulverizer is very suitable.

73. Ores of this class containing talc, or lime, require the addition, during or before the roasting, of from one to five per cent. of salt, to convert those substances into chlorides, otherwise they would consume a great deal of gas in the subsequent chlorination of the gold.

74. The furnace, if new, is dried during five or six days with the aid of a very small fire, after which, or

if it has been used before, it is heated for from eight to twelve hours, with a gradually increasing fire, but, for this class of ore, need not be red hot before charging, because there is so much sulphur in the ore that it soon ignites, and assists in heating the furnace. While heating the furnace, a quantity of the ore is dried on the drier, being turned over occasionally with a garden hoe, and a charge is put into the hopper.

75. Many operators put too much ore in the furnace, which increases the work of the roasters, without any corresponding advantage, because the term of roasting is nearly proportional to the thickness of the bed of ore on the hearth. I have found, by repeated experiments, that from ten to twelve pounds to the square foot of hearth is enough. The charge for the furnace described is, therefore, 1,000 pounds on each hearth, or a ton in all, so the charge for the hopper is half a ton.

76. When the furnace is ready, a charge of ore is dropped in on the second hearth, and at once moved, with hoe and spade, to the first, and there spread evenly. Meanwhile, a second charge is put into the hopper, dropped on the second hearth and spread. Care must be taken that none of this charge falls on the first hearth, for which reason the second hearth, as may be seen by referring to Plates 1 and 2, is made a foot longer, and the first foot has a slight downward inclination, as it is intended to be left bare of ore during the roasting, and to act as a partition between the two charges.

This is very important in roasting ore for chlorination, because the admixture of a little raw ore with that which is half roasted causes great delay in finishing; or, half roasted ore mixed with that which is roasted is injurious in the chlorination. Some operators build the hearths on the same level, and separate them by a low wall, over which the ore is lifted with the spade when transferred from the second hearth to the first.

77. *Roasting.*—When the sulphur begins to burn on the first hearth, the fire is kept quite low. The burning of the sulphur can be seen, even in the day time, if the furnace house is dark, as it should be, with no cracks or windows through which the sun can shine, a point of some importance, which is not sufficiently attended to. It is difficult to properly regulate the heat in full day light, but it is even worse if the sun shines in through chinks or windows, so that an open shed is better than an imperfectly inclosed room.

78. Continuous stirring of the ore is useless, for there is a certain quantity of sulphur to be burned, which requires a certain quantity of air, and there is no use in exposing a fresh surface until that exposed is burned, as far as can be done with the moderate heat allowed at this stage of roasting. The guide is, that the ore must be stirred as soon as it is seen to become dark on the surface.

79. The draft is regulated so as to cause as much air to enter as is consistent with the required heat; hence the doors on the working side are left open during all the earlier part of the roasting, though it may be necessary to close them during the intervals of stirring, at least on the first hearth, for the finishing at a high heat. Even the back doors may be left ajar at the rear ends, at times, with great advantage, because the ma-

terial now under consideration, being nearly half sulphur, requires an enormous quantity of air, without
which it cannot possibly burn, and the more rapidly
this is supplied the sooner will the oxidation be completed, bearing in mind, of course, that the proper degree of heat must be maintained. At the same time
it must be remembered that a high heat at first is an
actual disadvantage; the ore roasts faster and better
with a low heat, until oxidation is almost complete.

80. In thus starting a furnace, the roasting of the
charge on the second hearth will not make much progress until that on the first is so far advanced that a
higher heat may be used, when it also will begin to burn,
and must be stirred regularly, so that by the time the
first is finished, probably in twenty hours, it will be half
roasted, or more, and, when moved to the first hearth,
will bear a good heat. The roasting is continued, under a gentle heat, as long as the sulphur burns actively,
and when a blue flame, or a glow, is no longer perceived
on stirring the ore, the heat is gradually increased until
it reaches a light red, approaching yellow, if the ore
will bear it without melting into lumps, as it will if it
consists chiefly of iron or arsenical pyrites.

81. From what has been said about the burning of
sulphur, it is evident that the stirring should be done at
progressively shorter intervals until oxidation is completed. Care must be taken to stir quite to the bottom,
and in every corner; and it is advantageous to clear the
hearth entirely, a part at a time, and leave it so for a
few minutes to allow of the oxidation of the thin strat-
um of ore which is not moved by the hoe. The stirring is mainly done from the front of the furnace, with

the long hoe; but occasionally the roaster goes to the back, and with a small hoe explores the sides and corners within his reach, raking the ore out, and pushing it toward the middle of the hearth. In ceasing for a time to stir the ore, it is better to leave the surface ridged across by drawing the corner of the hoe over it, because the wavy outline thus given exposes a larger surface for oxidation, or for heating. This, as remarked by Kustel in his valuable work on this subject, is important enough to be attended to.

82. As the heat, in this kind of furnace, is unavoidably greater near the fire than at other parts, the ore must be changed from end to end, so that every particle may, at one time or another, be brought near to the fire-wall, or bridge. This is done by drawing it with the hoe into a ridge along the middle of the hearth, or better, diagonally across it ; then, with the spade resting on the roller bar as a fulcrum, moving that ore which is farthest from the fire, near to it, placing it on the empty space on the hearth, and the reverse. As long as sparks can be seen on flirting the ore with the hoe, or pouring it off the spade, the roasting is not perfect, nor does the ceasing of this appearance prove that it is so.

83. The roasting, under a strong heat, is continued until all the iron sulphate is decomposed, which may be known by taking a little of the ore on a small sampling shovel, and throwing it into a cup containing a little water. In a minute the water will be clear, and if, on adding a few drops of solution of potassium ferridcy-anide, or red prussiate of potash, a blue or green coloration is seen, the roasting is not perfect. After a little practice, this test may be dispensed with, as the iron

sulphate is very easily decomposed. The presence of a little of it is not fatal to the chlorination, but it causes a larger consumption of chlorine.

84. Copper sulphate, if present, remains after iron sulphate disappears, requiring a long time and a high heat for its complete decomposition, which, however, is not absolutely necessary. It gives, with the ferrid-cyanide, a brown or yellow precipitate, which, in the presence of iron sulphate makes a green coloration by mixing with the blue of the iron.

85. If it is desired to roast to complete decomposition of copper sulphate, a test is made by washing a small quantity of the ore·on a filter, in a glass funnel, with hot water. To the water which passes through the filter is added ammonia, and if even a very little copper is present, a blue color is produced. It will not do, in this test, to add the ammonia without removing the ore by filtration, because the blue color would then be produced by copper oxide, which remains after the sulphate is decomposed.

86. When the roasting is finished, a careful panning or horning of a little of the ore will, in general, show the gold plainly ; if not, then grinding finely in a mortar, and panning, should make it visible, at least with a lens. But it happens with some ore that very little of the gold can be seen even in this way, perhaps because of its extreme fineness, causing it to be lost in the most careful washing. No sulphides should be visible, or only a few minute particles which the grinding may have brought out.

The roasting of the first charge being completed, it is drawn from the furnace and spread on the cooling floor. The second charge is moved down to the first hearth, and a third is put on the second hearth. The charge now on the first hearth, having been all the time roasting on the second, will, if it has been properly attended to, be almost done, and, with a good heat, will be finished in a few hours, while the third, having consequently but a short time to roast on the second hearth, will take longer on the first, when moved down, but this irregularity will gradually disappear, and after a few days the charges, if equal in quantity and properly attended to, will remain an equal length of time on each hearth, and the discharges will take place regularly.

87. A furnace with two hearths, or even with three, does not allow of a continuous strong fire. The heat is reduced each time that a charge is moved down, and not until a strong smell of burning sulphur is no longer evolved by the roasting ore, is the heat raised gradually to the finishing point. The great fault of most beginners is in using too much heat, especially in the early stages of roasting. A waste of fuel and of time is caused by using a higher heat than is necessary, for the oxidation proceeds faster under a low heat at first, and the ore assumes a better condition for the action of chlorine, being more porous, and lighter. Especially is this the case when lead is present. Even at the finish, too high a heat may melt the gold particles together, or to a spherical form, which is disadvantageous, because of exposing a smaller surface than any other for the chlorine to act upon. The workman must use judgment, and acquire experience, before he can roast well, and with the least consumption of fuel.

88. The roasting of concentrated sulphides, consist-

ing chiefly of iron or arsenical pyrites, requires from 20
to 30 hours, in a reveberatory furnace with two or three
hearths.

89. The roasted and partially cooled ore is moistened, by spraying with water from a hose, the finger
being placed partly over the nozzle to spread the
stream, or a sprinkler may be employed. To equalize
the moisture, the ore is mixed by means of a hoe or
shovel. The degree of moisture required varies. If
the gold is in very fine particles, the ore is made only
so damp as not to dust, and to cohere slightly on being
compressed in the hand. If the gold is coarser, more
moisture is required, sometimes as much as can be used
without making the ore so wet as to settle into a compact mass, for it must be loose and porous in the chlorinating vats. The use of more water than is necessary is disadvantageous, because it causes a greater
consumption of chlorine.

It is beneficial to allow the ore to remain a number
of hours, even a day or two, in a moist heap before pro-
ceeding further, as this enables the moisture to pene-
trate and soften any lumps which may have been
formed in the roasting. If this is done, the surface of
the heap should be sprayed from time to time, to coun-
teract drying, and before charging a vat the ore must
be re-examined, to insure its being in the right condition
as to moisture. It is next to be placed in the vat.

90. *Charging the Vat.* If the filter in the vat is
made of sand, on the gravel, and is wet from previous
use, it is advisable to dry it somewhat by laying on it a
few hundred pounds of dry roasted ore, which soon becomes moist by drawing the water out of the sand,

whence it will be seen that, if the moistened ore were put in it at once, it would become too wet, and, by packing, would obstruct the passage of the chlorine, causing a back pressure which would force the gas through the water joints of the generator. After the dry ore has been an hour or so in the vat, and has become damp, it is removed, and the charging is proceeded with. If only gravel and burlap are used on the perforated false bottom, for a filter, this precaution is not needed, and time is saved.

91. When charging a vat for chlorination, the discharge pipe is stopped with an oiled plug, and hung up, and the moist ore is sifted into the vat. The lumps which are retained by the sieve, are removed from time to time, to be afterwards recrushed, with rollers or otherwise, and returned to the roasting furnace. Any scraps of iron which may have found their way into the ore, and which would be very detrimental to the chlorination, are removed by the sifting. As the ore cannot be spread evenly over the filter by the sifting, and as it packs a little where it falls, it is distributed and loosened with an iron rake. The vat is filled to within four inches of the top, and the ore is very slightly packed around the side, by pressing it with the hand, so that the chlorine cannot make its way between it and the staves, but is forced to permeate the entire mass.

92. *Chlorination.*—Time may be saved by starting the chlorine generator as soon as the vat is about half charged, as the filling will be completed before the gas can reach the surface.

The progress of the chlorine is watched, after the charge of ore is in, by making a few little excavation and holding therein a glass rod dipped in ammonia, which instantly shows the presence of chlorine by a dense white fume. The moist stopper of the ammonia bottle may be used instead of a glass rod. As soon as the chlorine has risen to within a few inches of the surface of the ore, the holes are lightly filled, and the cover is let down into its groove, or rabbet, and luted with a paste made of clay and sand, in such proportions as not to crack in drying; or the luting is kept moist by placing wet cloths on it. The plug hole in the cover is left open until the vat is quite filled with chlorine, as shown by the fume when ammonia is held to the hole, which is then tightly plugged.

If at this time the chlorine generator is not required for another operation, and if the evolution of gas has become so slow that it only bubbles through the wash bottle at intervals of many minutes, it is well to leave it connected with the vat; otherwise it is disconnected, the lead nipple in the vat is plugged, or the vent pipe connected, and the chlorine pipe is either applied to another vat, or the end is passed out of doors. The vat is now left undisturbed for from 12 to 40 hours, to allow the chlorine to act on the gold.

When two or more vats are impregnated at one time, the forked lead pipe is used, the flow of chlorine to each vat being controlled by a pinch-cock, for which a clothes pin answers very well, applied on the rubber part of the branch pipe; or a lead pipe may be laid alongside of the vats, with a nipple and lead faucet to connect with each of them by a short piece of rubber tube.

93. Leaks in any part of the apparatus are easily located by the aid of ammonia, as above explained, and

the operator's nose should be excused from this duty, or serious consequences may follow. Leaks in pipe-joints can usually be stopped with oil or lard, and those in the vat-cover with a paste of oil and sand, or earth. The charging of a vat accupies from one to two hours, as the sifting is very tedious.

94. *Making Chlorine.* To produce enough chlorine for three or four tons of roasted ore, the charge for the generator is about:

Manganese............................ 30 pounds
Salt................................. 40 "
Water................................ 36 "
Sulphuric acid of 66° Beaumé.......... 70 "
Or the same of salt and manganese,
Water................................ 32 "
Acid of 63° Beaumé................... 74 "

The cleaning out pipe of the generator being securely plugged, the salt, manganese, and water may be put in at any time, and stirred with a wooden paddle. The acid is weighed or measured, and kept at hand in a vessel of lead, or enamelled iron, though the enamel generally soon peels off. The naked iron does very well for strong acid. The cover of the generator being let down, the water joints all filled, and the pipes connected, a fire is made in the fire-place, or steam is turned into the heater, to warm the apparatus some time before the chlorine is wanted.

When all is ready for the chlorine to enter the vat, a portion of the acid is poured into the generator through the funnel, and mixed by turning the stirrer. Chlorine soon begins to pass, and can be seen in the wash bottle by its greenish color. It must pass in a lively stream through the water, and when the current slackens, more acid is put in until the whole is used.

From time to time the stirrer is turned carefully. Sometimes it must be lifted a couple of inches before the impacted stuff can be loosened, and this is the reason why the water joint in this part is made two inches deeper than in the others ; otherwise the lifting of the stirrer might cause an escape of gas which would be injurious, and might be dangerous, to the workman.   During the operation the apparatus is gradually heated, but not so as to make the contents boil.

By the time the charge is exhausted the test by means of ammonia should indicate an abundance of chlorine issuing from the holes in the vat covers.   If this is not the case, more manganese and salt must be put into the generator, through the opening provided for that purpose, but, as the workman is liable to suffer serious inconvenience from the escaping gas while doing this, it is better to avoid the necessity by putting in enough at first.   If an additional quantity of the other materials is used, more acid will also be required, unless an excess was taken for the charge, but this can be introduced without inconvenience by means of the pipe and funnel.   The relative proportions of the materials vary according to their purity, and the quantity of chlorine required depends much on the character, and more or less perfect roasting, of the ore ; hence no fixed rule can be given for either.

95.   When it is desired to prepare the generator for another operation, the end of the rubber tnbe leading

from the wash bottle is passed out of doors, or connected with a charged vat. Water is poured into the water chamber, *b*, and passes to the interior of the generator through notches cut in the lower edge of the cover for that purpose, and fills it, expelling all gaseous chlorine, which would otherwise escape into the room on raising the cover, or removing the plug at *d*. The water is allowed to stand for some time in the generator, to dissolve the sodium sulphate formed by the action of sulphuric acid on salt, which sometimes forms a hard incrustation, if too much heat has been applied. The plug at *d* is then withdrawn, and the generator is emptied through the pipe, which should deliver out of the room.

If there is any impacted residue which cannot be removed by a stream of water from a hose, directed through one of the openings, the main cover is raised and the stuff carefully scraped off the bottom, but if not, the recharging can be done by the opening, *h*, without raising the cover. If the residue contains much unchanged salt, the proportion used should be reduced. An excess of manganese shows itself by its blackness. When the ingredients have been correctly proportioned, only a dark grey sediment remains with the more or less acid liquid.

96. *Leaching the Gold Ore.*—After the ore is supposed to have remained long enough in contact with chlorine—that is, from 12 to 40 hours—the plug hole in the cover is opened, and tried with ammonia as before. This must still produce a dense fume; if not, the chances are that a poor result will be obtained in leaching, unless chlorination is repeated. If appearances

4

are satisfactory, a little ore is taken out through the
hole with a sampler, finely ground, and carefully washed
in a saucer, or a horn.   Neither gold nor sulphides
should be visible with a lens, but, as before remarked,
some ores do not show the gold well, even before chlorination, and for such this test is not conclusive.

97.   At this point a difficulty arises, from the fact
that a large quantity of chlorine remains in the vat, the
escape of which into the leaching room is not desirable.
If the cover of the vat be raised, the gas is visible like
a green sea above the ore ; a pestilential sea whose
waves, surging forth upon the slightest disturbance of
the air, threaten to envelope and suffocate the workman, unless he hold his breath while hoisting the cover
and fastening the tackle, and then retreat in haste till
the storm is over ; and even then the same trouble re-
curs while water enters, and displaces the chlorine still
remaining beneath the filter, and permeating the loose
mass of damp ore.

To overcome this difficulty, some operators fill the
vat with water before lifting the cover, and, by means
of a rubber hose, convey the expelled gas, either out of
doors, or into another vat charged with ore.   There
are two objections to this plan—firstly, a great deal of
chlorine is absorbed by the water, which has an un-
favorable effect in the precipitation of gold, making it
more difficult to settle, as well as wasting the precipitant ; and, secondly, the water, entering the tub in one
large stream, descends through the bed of ore at one
point, and rises through it in other parts, giving a solution of nearly uniform strength throughout the mass,
and requiring a large quantity of water to complete the

leaching, taking more time, and giving finally a weaker solution than is necessary, for the precipitation takes place better when the solution is strong.

The best way to apply the water is by spraying it all over the surface of the ore in the vat, so that, percolating downward through the mass, it carries the greater part of the gold chloride in the first portion of the water, and a rich solution is obtained at once, while less water is required to remove the whole of the gold, and the stronger solution thus finally obtained gives a better precipitation on addition of iron sulphate. But if the spraying is undertaken by hand with a sprinkler, the loose ore, settling as it becomes wet, expels chlorine in such volumes as to render it impossible for a man to stand near the vat.

The course I adopted was to place a coil of lead or rubber pipe, in which were numerous small holes, around the vat under the cover. After raising the cover, the water supply pipe was connected with the perforated coil, and the water, under moderate pressure, issued in a number of fine jets, and showered all over the ore, through which it descended gradually, dissolving the gold chloride as it passed, so that the first solution which passed through the filter, and filled the space under it, was very rich, while that which finally remained on the top was very weak. The discharge was then started, and the water thenceforth admitted in the usual way, keeping the surface of the ore covered until the leaching was finished. This plan, however, is still open to the objection that it gives a solution containing much free chlorine, while the chlorine which is expelled is not removed from the room; hence it is very desirable that means should be used to abstract the surplus from the vat before raising the cover, or admitting water. The appliance I am about to describe will not only do this, but will also save the chlorine, for use in another charge of ore, with little waste.

98.   Before lifting the cover, the plug-hole being open, one of the lower openings, either the nipple through which the gas was admitted, or the discharging pipe, is connected by a rubber hose with the interior of an inverted bell of sufficient size, which is immersed in a tank of water, precisely like the "gasometer" of gas-works, and like it, counterpoised and suspended by ropes or chains passing over rollers, so as to allow the bell to rise or fall in its water tank.   By adding weights to the counterpoise, the bell is gradually raised, and the chlorine is withdrawn from the vat and enters the bell, while air, entering the vat through the open plug-hole in the cover, replaces the abstracted chlorine. The filling of the vat with water is then proceeded with in any way that may be convenient, without annoyance from escaping chlorine.   The chlorine in the bell may be again expelled when wanted, and caused to enter the same, or another vat, by simply removing a part of the counterpoise, and allowing the bell to descend by its own weight.

When the ore in the vat is covered by the water, the discharge pipe is let down to the trough which leads to the gold tub, and the plug is removed from it.   To prevent disturbance of the ore by the water flowing into the vat, the stream is received on a perforated board, or on a sack laid upon the ore.   The inflow and the outflow must be equal, so that the ore remains submerged. If the outflow is too rapid, it is lessened by simply " kinking " the pipe, which, for this reason, is of pure rubber, and long enough not only to reach the trough, but a foot or so more ; it can then be "kinked " so as to entirely stop the flow if desired.

99. A few ounces of the strong solution of gold chloride first obtained, are set aside for a purpose which will shortly appear. If the ore contains no copper, and has been properly roasted, the lixivium will be of an amber color. A green hue indicates the presence of copper, and a very dark, almost black, appearance, produced by iron perchloride, shows that the ore has been improperly roasted.

100. *Precipitating the Gold.* — On beginning the leaching, about a dozen gallons of the solution of iron sulphate is let into the gold tub (for exceptions see article on precipitating gold, under " Addenda.") The gold is precipitated as a brown powder, which, however, requires many hours to settle. The object of putting the precipitant into the gold tub in advance, is to decompose the gold solution, and precipitate the gold, as soon as it enters the tub, so as to lessen the loss by absorption into the wood, by an accidental leak, or an overflow ; also to neutralize the free chlorine in the lixivium.

101. When the solution flowing from the leaching vat has become colorless, or before, if much copper is present, a little of it is received from the hose in a glass vessel, such as a beaker, tumbler, or ore sample bottle, and some solution of iron sulphate is added. If gold is present, a dark cloud is produced, either instantly or after the lapse of a few seconds. When only a

slight discoloration is produced in the lixivium, by the addition of iron sulphate, it is better to divert the stream into one of the extra gold tubs, or "gold wash-tubs," because the metal obtained from very weak lixivium is so extremely fine as to settle with difficulty, and even to pass through a filter. The leaching is continued until not a trace of gold can be detected in the lixivium coming from the vat.

102. I will here call attention to a circumstance which was observed in my works, After the leach ceased to show gold by the iron sulphate test, if the discharge was stopped for a few minutes, and again started, the solution would again give a considerable precipitation on being tested, but in a few minutes would again fail to show even a trace, and this could be repeated several times. It was due to unequal percolation, and is mentioned here in case the same might occur in other works.

103. The solution in the gold tub is well stirred with a wooden paddle, and, after waiting for a few minutes, to allow the gold to settle a little, a sample is taken in the glass, and tested by adding some iron sulphate. If any discoloration is produced, more iron sulphate is needed in the tub. If none, yet, as an excess does no harm, it is best to make sure that there is enough of the precipitant, by emptying the glass, rinsing it once or twice with the liquid from the gold tub, taking another sample, and testing it by adding some of the strong solution of gold, which, it will be remembered, was set aside at the beginning of the leaching. If a dark cloud is now produced, it is certain that enough of the iron sulphate has been used in the gold tub. If

not, it is safest to add another bucketful, mix thoroughly, and try again. By taking the trouble to make this check test, an expensive accident may be avoided.

104. The gold settles better if the stirring is repeated after an hour or two, and it is advantageous to add a few pounds of sulphuric acid. If, in stirring, a circular motion is imparted to the liquid, nearly all the gold will settle near the middle of the tub. The gold requires from 24 to 48 hours to settle.

105. The weak leach in the wash-tub may be treated with some of the iron sulphate, to throw down the small quantity of gold which it contains, which must then be allowed abundant time for settling; or it may be raised by means of a wooden or leaden pump, and used to begin the leaching of the next charge of ore. A third way, which suggests itself, but has not been tried, except, as will be seen hereafter, in connection with the extraction of silver, is to precipitate the gold, together with any remains of copper, etc., which may be present, by an addition of calcium polysulphide, the preparation of which will be described in its place. A great excess of the precipitant should be avoided, because the gold tersulphide which it produces is somewhat soluble in calcium polysulphide. The precipitate obtained in this way would probably settle more rapidly than the other.

106. *Iron Sulphate Solution.* This precipitant for gold may be prepared by dissolving common copperas, or green vitriol, in water. A quantity of the crystals is placed in the vat prepared for it, which, it will be remembered, contains a filter, and water is added. There should always be an excess of copperas on the filter, in order that the solution may be saturated. It may also be

prepared by dissolving scrap iron in diluted sulphuric acid. Several days must be allowed for the preparation of the solution by this method, and an excess of iron must always be present in the dissolving tub, the supply of solution being maintained by the addition of dilute acid from day to day, and of iron as fast as it disappears.

107. When the liquid in the gold tub has become perfectly clear, the discharge pipe is freed, with its float, and the liquid is allowed to flow through a filter of sand or sawdust, because some fine gold always remains in suspension, even after 48 hours' settling. If a sand filter is used, it is, after a time, taken up and chlorinated, like ore. If if is of sawdust, it is thrown into the furnace with ore, and burned, or burned by itself and the ashes collected.

If the discharge pipe of the gold tub does not sink deeply enough to draw, a piece of board is laid over it, which keeps it below the surface, and follows it down. If it tends to draw too deeply, by being sucked down, a larger float should be used; or a cord attached to the float, and turned around a nail in the side of the tub, will secure it; but it is then necessary to slacken the cord from time to time. If the hose is rather stiff, and well adjusted, it will give no trouble. The construction of the float, encircling the end of the hose with a thickness of two inches of wood, obviates all danger of a loss of the precipitated gold by drawing too near the bottom of the vat. Other arrangements for drawing off the liquid will be found under " Addenda."

108. *Collecting the Gold.*—Unless when working very rich ore, or for special reasons, the gold is not re-

moved from the tub until several precipitations have been made. When a clean-up is desired, the liquid is drawn off as closely as may be done without loss of gold as described; the remainder, with some of the gold, is allowed to flow through the faucet into a small tub. The greater part of the gold remains on the bottom of the precipitating tub, and is taken up with a scoop, and carried in a suitable vessel, such as an enamelled kettle, to the filters. The bottom of the gold tub is washed by means of a small stream of water from a hose, and a whisk broom. The washings are added to the drainings in the small tub, and are either allowed to settle again, and again drawn off, or taken at once to the filters.

109. *Washing the Gold.*—When drained in the filters, the gold, if nearly clean, which is known by its having a brown color, not too dark, may be washed at once on the filter, with hot dilute sulphuric acid and salt, or dilute hydrochloric acid, and afterwards with hot water alone; if very impure, it is transferred to a large porcelain dish, and boiled on a sand bath with slightly diluted acid. After adding water, it is again thrown on the filter, drained, and washed with hot water until the latter comes through tasteless. The filter bags, to which some gold adheres, are kept under water in a suitable vessel, to prevent rotting, until wanted again. When worn out, they are lightly sprinkled with powdered nitre, dried in an iron dish, and touched with a live coal. They then burn like touchpaper, and the ashes are added to the precipitated gold. Too much nitre must not be put on the bags,

or the combustion will be too violent, and gold may be lost.

110. *Drying the Gold.*—The washed gold is pressed by hand in the filter bags, transferred to an iron dish, and almost, but not completely, dried. While drying it is mixed with a little powdered borax and nitre. If several ounces of nitre are used, on account of impurities in the gold, it is advisable to add about half as much clean quartz sand, to protect the crucible from the action of the nitre during the subsequent melting of the metal. The purpose of allowing the pulverulent metal to retain a little moisture is, the prevention of loss by dusting while putting it into the melting pot.

111. *Melting the Gold.*—If the gold is very clean, it may be melted in a black lead pot, but I have always preferred a sand pot, on account of the tendency of plumbago to cause contamination of the gold by reduction of the base metal compounds which may be present. There is, however, some slight risk of the sand pot breaking, nothwithstanding the precaution of adding sand with the nitre, and well drying the pot before putting it into a moderate fire. To guard against being obliged to wash the gold out of the ashes in case of breakage, with some risk of loss if the bottom of the ash-pit is rough, a black lead dish, made of the lower part of a crucible, is placed under the sand pot.

112. The fuel used for melting is charcoal or coke. English coke is so much better than that from the gas works as to compensate for the extra cost. A common portable assay furnace is very suitable for the melting of moderate quantities of gold, and a sand pot about ten inches high is large enough for several thousand

dollars worth, although a larger crucible is needed than for metal which is in a more compact form.

113. When the crucible is red hot, it is taken from the furnace and gently tapped, to ascertain that it is sound, then filled with the gold, by means of a scoop, replaced in the furnace, covered, and gradually brought to a strong red heat. As the partly melted gold settles, the pot is refilled, without removal from the furnace, by means of the scoop, and, if required, a sheet iron funnel. It is important to refill the pot before the contents are quite fused, and while they are in a pasty condition; otherwise some of the gold may be thrown out of the crucible. The handling and melting of the gold are greatly facilitated if the precipitated metal is pressed into cakes, and heated to redness in a muffle, as is done in the mint with the finely divided gold from the refining. When the entire charge is in, a white heat is maintained until the gold and slag are thoroughly melted. An addition of more borax may be required if the slag is not sufficiently fluid. The mass is stirred with a red-hot iron rod, which must not be kept in too long, or it will make the gold base. Some metallurgists use a strip of black lead for the purpose.

114. When the slag is quite liquid, it is skimmed off the melted gold, by means of a piece of nail rod, which is turned at one end to a flat spiral, and bent to a suitable angle for convenient use. This is dipped, cold, into the melted slag, quickly withdrawn, and pressed upon a cold block of greased iron, so as to flatten the adhering slag, then just dipped into cold water, and again into the slag, repeating the routine until the latter is all removed, and the melted gold is seen in the pot.

115. It sometimes happens that the precipitated gold is contaminated by lead sulphate, which cannot readily be removed in the washing, and which, does not mix with the slag proper. This cannot be skimmed off in the manner described, but is removed by means of a red-hot scorifier held in the tongs.

116. If the melted metal is pure, it appears greenish and motionless, but if any base metal remains, colored rings, or spots, are seen, moving from the center outwards. These rings are base metal oxides, and if they are supposed to be caused by iron, or copper, a few pieces of borax are thrown in, and allowed to melt while the pot is left uncovered. After a time the new slag thus formed is skimmed off, and more borax put in, and, if the metal is very impure, a little nitre is added to help the oxidation of the base metal, for if the operation takes too long a time, the pot will be cut immediately above the gold. If the impurity consists of lead, a little bone ash is better than borax to absorb the oxide, as it is formed, especially when a black lead pot is used.

117. When all is ready, the pot is seized by the tongs and the metal is poured into a cast-iron ingot mould, which has previously been warmed and smoked, or oiled. A little oil is poured on the top of the bar before it solidifies. Unless the gold is to be assayed and stamped at the works, it is not necessary to be careful about the shape or appearance of the bar. It is then as well, and saves some trouble, to allow the metal to cool in the pot, which is then broken; in any case it can be used but once. But this cannot be done if a black lead pot is used, because it will serve many

times.   The slag and pots are preserved, and either sold, or crushed and treated with quicksilver for the gold which they contain.

118.   *B. Concentrated Pyrites Containing Gold and Silver.*—This material is obtained in the same manner as the concentrations containing gold only.   The silver is usually of secondary importance as to its value.

119.   *Roasting.* — This is conducted in the same manner as that for gold, but as the chlorination of the silver in the vats, by means of cold chlorine, does not give satisfactory results, it is better to form the silver chloride in the furnace, for which purpose an addition of salt is necessary.   One per cent., or 20 pounds to the ton of ore, is generally enough for fifty ounces of silver. In some cases the salt is mixed with the ore when charged, but this is not always safe, as a very large loss of gold sometimes results.

120.   If it is found, by means of assays, that a loss results from charging the salt with the ore, the course adopted is, to roast as directed for gold, then to let the ore cool somewhat in the furnace, and throw in the salt through one of the doors, scattering it as much as possible over the ore.   The doors are then closed until the salt ceases to crackle, after which it is rapidly and vigorously mixed with the ore by means of the hoe, and the charge is drawn out within 20 minutes after putting in the salt.   The heat must be high enough to cause the reaction between salt the and the metal sulphates, yet not so high as to produce heavy fumes, or blue flames ; nor should a yellow substance be seen on the side of the furnace, where the ore is drawn out. The hot ore evolves copious fumes of volatile chlorides

and free chlorine. It is not spread at once, but is left in the heap to be acted upon by the gases for an hour, after which it is spread on the cooling floor.

121. The best results are produced by adding to the ore, before roasting, a little more salt than the quantity required by theory to chloridise the silver present; that is, for fifty ounces of silver not more than two or three pounds of salt, which should be finely ground. The one per cent. should be added at the finish as above. The use of a little salt at the beginning in this way caused no blue flames of copper chlorides, heavy fumes, nor loss of gold, while it caused an increase in the yield of silver amounting to three or four ounces to the ton. It was not proved whether the result would have been the same if the final addition of one per cent of salt had been omitted, as it could do no harm, and cost almost nothing. But when no salt whatever was used, the result was unsatisfactory, and considerably more chlorine was required in the vats.

Ores differ much in their behavior in working, and without the cause of the difference being manifest, even from an analysis. Thus there are cases in which ten per cent of salt has been mixed with the ore before roasting, without causing a loss of gold. It therefore behooves the operator not to rely solely on the instructions given in any book, but to watch his results carefully and constantly, until he has established a satisfactory mode of procedure.

122. After roasting, the work is carried on precisely as directed for ore containing no silver, up to the point at which the gold leaching is finished, except that it is not necessary to extract the last trace of soluble gold

with water, because, if a little remains, it will be obtained with the silver. It is, however, proper to remark that, if the ore is of such a character that much salt must not be used at the commencement of the roasting, the lumps from the sifting should be washed, to remove soluble chlorides, before being re-roasted.

123. *Leaching the Silver.*—At this point some operators remove the ore to other leaching vats, from an idea that the small quantity of the leaching liquid used for silver which remains in the filter, is injurious to the chlorination of the next lot of ore. I do not think it worth while to take this trouble. I even found it unnecessary to take any special pains to wash the filters before recharging, any further than is done in washing the hypo out of the ore, which is necessary in order to avoid waste. The traces of hypo remaining, if not oxidized to harmless sulphate by the action of air, must be so, instantly, on the admission of the chlorine. It is true that in the latter case hydrochloric acid would be formed, but in such minute quantity as to be of no importance. At the worst the only inconvenience would be a slightly increased consumption of chlorine.

When the gold leaching has been carried as far as is desired, the water in the vat is allowed to subside below the surface of the ore. The solution of calcium hyposulphite is then admitted, and the discharge is directed into one of the extra silver or wash tubs.

It is best to have a separate trough to convey the silver solution to the tubs, not only because it is convenient to have the gold and silver tubs at opposite ends of the set of leaching vats, but also because it is often necessary, when working several vats, to run both gold and silver solution at the same time. I therefore have two troughs, extending in front of and below the leaching vats, side by side, but inclined in opposite directions, so as to lead, the one to the gold, the other to the silver tubs.

The hypo should never be allowed to flow into the gold tubs, because it forms a combination from which the gold cannot be precipitated by iron sulphate. As, however, the precipitant used for silver throws down gold also, whether dissolved in water or in hypo, there is not the least danger of losing any gold which may find its way to the silver tubs.

124. As the water runs out of the ore mass in the vat, the hypo follows it, and as soon as a *sweet taste*, indicating the presence of silver, is perceptible, the stream is turned into the main silver tub, because the wash water, after precipitation of any metal it may contain, is thrown away, but the hypo is preserved for future use. The flow is not allowed to be too rapid, but is checked by "kinking" the pipe, as directed in the gold leaching. There is no guide for this but experience, and the strength of the silver solution coming from the ore, which is known by its more or less sweet taste.

125. When a sweet taste is no longer perceived in the solution coming from the vat, a test is made by taking some of it in a glass, and adding some solution of calcium sulphide. If a precipitate is produced, it is certain that metal of some sort is still being extracted from the ore. The question then is, what metal, or rather whether there is any, or much, silver, for it is an unfortunate circumstance that the hypo dissolves lead sulphate, and other base metal compounds, es-

pecially after the extraction of silver is finished, or when the hypo is too strong.

Some idea of the nature of the metal which is being extracted by the hypo, may be obtained from the color of the precipitate. Silver gives a dark brown, copper a reddish brown, lead and antimony light, sometimes yellowish shades. It is, however, often impossible to say, from the appearance of the precipitate, whether it contains silver or not, and the sweet taste, when it is no longer strong, may be masked by other and less agreeable flavors. I have not been able to find a very simple test for the presence of silver in this case, but I will give the one I use, which can be made in ten minutes.

126. The precipitate in the glass is stirred, or shaken, to make it curdle, allowed to settle for a few seconds, and a portion of the liquid is poured off; the remainder, with the precipitate, is heated in a small porcelain dish. The precipitate blackens, shrinks, and settles, so that nearly all of the liquor can be poured off without much loss of precipitate. A little nitric acid, and a fraction of a grain of salt, are then added, and heat is again applied. The action is very rapid, and in a few seconds nothing remains of the black precipitate but a yellow mass, principally sulphur, but possibly in part silver chloride. To prove this, ammonia is added, very cautiously unless after cooling and diluting, until the contents of the dish smell strongly of it.

If copper is present, it gives a blue color; silver chloride is dissolved in the ammonia, which is then poured upon a filter, and to the liquid which passes

5

through, and is received in a test tube, nitric acid is added, drop by drop, while the test tube is inclined away from the operator's eyes, until the smell of ammonia is no longer perceptible. If there was any silver in the precipitate, it now appears as a white cloud of chloride, which curdles on being shaken, or, if in very small quantity, as a slight milkiness in the liquid. A little practice will enable the operator to judge as to whether it is worth while to continue the leaching, or not.

127. When the quantity of silver in the leach is inconsiderable, the hypo is turned off, but, as it is not to be wasted, water is again led into the vat as soon as the ore is uncovered, to displace the hypo which would otherwise be retained. As soon as it is found, by the taste, that the hypo is almost washed out, and nearly pure water begins to come, the stream is turned into one of the wash-tubs, and the influx of water to the vat is stopped. The ore is allowed to drain, to facilitate which it may be dug over with a shovel. It is then removed.

128. The leaching of a charge occupies from eight to forty-eight hours, according to the richness and character of the ore, and the skill displayed in the roasting.

129. *Precipitation of Silver.*—The silver is precipitated by means of a strong solution of calcium polysulphide, called, in the works, simply "calcium," or "sulphide," which is led into the silver tub, by a hose from the elevated vat in which it is kept. It throws down the silver, and other metals, as sulphides, in the form of a dark brown mud, which soon turns black.

At the same time it restores the hypo, which was altered in dissolving metals, so that it can be used again.

It is imperative that not more of the calcium sulphide be used than the quantity required to precipitate the metals.   If a little silver remains in the solution, it is not lost, as the hypo is used again; therefore, it is best to add *a little* less of the calcium sulphide than would be required to precipitate the whole of the metal.   If too much is used, the excess remains unchanged, and mixed with the hypo, in the re-use of which it converts some of the silver chloride in the ore into sulphide, which cannot be leached out.

In practice it is not difficult to precipitate aright.   A circular motion is given to the solution in the silver tub, the stream of calcium sulphide is turned in, and allowed to run as long as it is seen to cause a distinct precipitation.   The liquor is then stirred, thoroughly and vigorously, for a couple of minutes.   The circular motion is checked by a reversed movement of the stirrer; a few minutes are allowed for partial settling, a sample is taken in a glass, and a little calcium sulphide added.   If a considerable precipitate is produced, a few more gallons of the precipitant are run into the tub, and the stirring and testing are repeated.   When the calcium sulphide produces only a very slight precipitation, the contents of the silver tub are left undisturbed for a few hours.

130.   If, in the test, calcium sulphide gives no precipitate, it may be that too much has been used in the tub; therefore another sample is taken, and tested by adding a few drops of solution of iron sulphate, which

instantly gives a black precipitate if there is the least excess of calcium sulphide. If neither of the tests gives a precipitate, all is right; but if it is found that too much calcium sulphide has been employed, it must be counteracted by an addition of silver solution from another vat, or, if there is none to be had, some iron sulphate may be used instead. There is, however, no trouble if care be used. The experienced workman knows almost the exact moment when enough calcium sulphide has been added, by a white cloudiness which appears in the liquor.

The precipitate settles in a few hours, and the renovated hypo is drawn off, by means of an arrangement similar to that used in the gold tub, and pumped into the elevated tank, whence it is again led to the leaching vats as required.

131. If, when the leaching of a vat is finished, a dark scum is seen on the surface of the ore, it is taken off and returned to the roasting. It is caused by a small quantity of metal sulphide, which remained suspended in the hypo when it was drawn from the silver tub, and is thus filtered out by the ore. It contains more or less silver.

132. The metal contained in the weak solution in the wash-tubs, is also precipitated by means of calcium sulphide, but, after the settling, the liquid, which is little more than water, is allowed to run to waste. The metal should be completely precipitated, because, as the liquid is not used again, any which might remain in solution would be lost. An excess of the precipitant can do no harm in this case.

133. *Collecting the Precipitate.*—After several pre-cipitations, or when desired, the black mud, consisting of silver sulphide, mixed with free sulphur, and more or less base metal sulphide, is run out through the clean-up pipe, faucet, or plug-hole, to a filter, drained, washed by passing hot water through it, again drained, pressed into cakes or not, as desired, and dried.

134. *Roasting the Precipitate.*—The dried precipi-tate is next roasted in the small reverberatory furnace, to burn off the greater part of the sulphur, beginning with only the heat required to set fire to it, and gradually increasing the temperature to dark redness, or to such a degree as the material will bear without melting. It must be stirred while roasting, and changed from end to end of the furnace, in the manner directed for the roasting of ore. If but little base metal is present, the roasting is continued until as little as possible of the sulphur remains; otherwise a prolonged roasting is avoided, on account of the formation of too much base metal sulphate and oxide, which are injurious to the black lead pot in the melting.

135. *Melting the Silver.*—This is done in black lead crucibles, in a wind furnace, with coke or charcoal for fuel. If the precipitate, before roasting, contained but little base metal sulphide, the silver is seen in the form of threads traversing the roasted mass, which, however, still retains a considerable quantity of sulphur. The crucible, containing some scrap iron, is filled, and placed in the fire, standing on a piece of firebrick; for as the melting occupies a considerable time, even a thick layer of the best coke does not last long enough to prevent the crucible from settling down to the grate.

A little borax is added, and the whole is heated till there is room in the pot for more material, when it is refilled by means of a scoop and funnel. As in the case of gold, the refilling is done before the mass in the crucible has become fluid, in order to avoid loss by projection. As fast as the scrap iron disappears, more is put in; but, if such addition is made after full fusion, the iron is first heated.   .

If the roasted precipitate contains much copper or iron, more borax is required, and a little clean sand is useful, especially if the roasting has been excessive. Some charcoal is also added.

When the pot is full of thoroughly melted matter and pieces of iron, a test is made by placing the red hot end of a piece of nailrod, or thick iron wire in it. If, after a few minutes, on withdrawing the rod, it is found that a part of it has been melted, more time must be allowed. When iron is no longer consumed, the melting of that quantity of precipitate is finished, and slag and matte are dipped out, by means of a red hot assay crucible held with the crooked tongs, and poured into a mould, or iron pan. The pot is now refilled with roasted precipitate, taking the precaution to add it slowly until the melted mass is somewhat chilled.

When all the precipitate has been thus worked up, or the pot contains a sufficient quantity of metal, a part of the slag and matte is removed as before, and the remainder, with the silver, is poured into a warmed and greased mould. The overflowing of the slag and matte is of no consequence, if the mould is large enough to contain the silver, which will go to the bottom in consequence of its greater specific gravity. After removal from the mould, it is usual to place the bar in a tub of water, for the purpose of cooling it, but when there is matte upon it this·must not be done until the matte also has solidified; otherwise an explosion will occur.

If the melting has been properly conducted, the matte is brittle, and separates readily from the cooled bar. If it is tough, that which adheres to the metal must be beaten off, and the whole remelted in presence of iron, as it then contains a great deal of silver. After cooling, it is broken and examined for any large buttons of silver which may have been dipped out with it. All the slag and matte is preserved, the former to be sold, the latter to be crushed and reworked by roasting and leaching. If a handsome bar is desired, it must be remelted with borax, cleaned by skimming, and re-cast, covering the surface with powdered charcoal before solidification.

136. *C.—Concentrations containing Silver, but little or no Gold.*—These concentrations generally contain less iron pyrites, and more lead, zinc, copper, antimony, and arsenic, which makes them more difficult to roast. They require from 5 to 20 per cent of salt, according to their richness. The salt is usually charged with the ore, but as some ores are more liable to melt if it is added at the commencement of the roasting, than if the addition is made at a later period, it is sometimes necessary to complete the oxidation, under a moderate heat before the salting.

137. *The roasting* is commenced with only the heat required for the ignition of the sulphur. Stirring is

kept up almost continuously, in order to prevent sintering, or partial melting. If the salt has not been mixed with the ore at the time of charging, it is added as soon as the smell of burning sulphur has become faint, under a dull red heat. The heat is then raised cautiously, the ore swells, becomes flaky, and somewhat sticky, appearing as though it were moist, which is caused by the fusing together of the metal sulphates and the salt.

At this period the ore must not be stirred too frequently, or more than is required to ensure an equal heat, as it facilitates the escape of chlorine, which should be retained as long as possible within the mass. The furnace doors may be closed while the stirring is intermitted, because what is now required is heat rather than air.

It is a good plan to heap the ore near to the fire-wall, and let it remain for a time, then spread it, and again gather it into a heap. The roaster should be careful to avoid pushing the somewhat adhesive ore against the wall of the furnace with the hoe. The charge on the finishing hearth must be changed from end to end at least twice; once before the heat is raised, and again after. It is drawn out while still giving off an abundance of chlorine and volatile chlorides, and is allowed to remain an hour before being spread on the cooling floor.

138. The degree of heat which may be used depends on the quality of the ore. Zincblende requires a high temperature for its decomposition. Iron, copper and antimony sulphides, especially the latter, require a low heat at first, but are not very liable to cause sinter-

ing after oxidation.   Lead is troublesome, as all its compounds are very fusible.

Some ores do not bear a high heat, after the addition of salt, without a great loss of silver; yet if zincblende is present a strong heat is requisite for its oxidation.   Such an ore is roasted without salt until a sample, ground in a mortar, and carefully horned, or panned, shows no sulphides.   It is then allowed to cool to a moderate red heat, salted and mixed.   If a strong odor of chlorine is not thus developed, some dried copperas, in the proportion of about sixty pounds to the ton of ore, must be added and mixed.   When the proper smell is obtained, the charge is withdrawn.

139.   *Washing the Ore.*—The roasted and cooled ore, previously sifted if lumpy, is placed in the leaching vat, and cold water is passed slowly upward through the mass, the discharge pipe having been connected with the water supply for that purpose.   When the vat is full, the water is admitted above the ore, and allowed to flow out through the discharge pipe.   The washings are allowed to run to waste, unless they contain copper, in which case they are conducted to vats which contain scrap iron, by which the copper is precipitated.

The reason for introducing the water at first in the manner described is, that a certain quantity of silver chloride is dissolved by the strong solution of base chlorides and residual salt, which is formed on the first introduction of water.   By operating as described, this solution is brought above the ore, and on being diluted with a further quantity of water, deposits the silver chloride in and upon the ore mass, where it re-

mains until the hypo is admitted, while the major part
of the base chlorides, remaining dissolved, is removed
by the washing.

A portion of base metal chlorides, however, is
also deposited with the silver chloride, and, on this
account, some operators prefer to admit the water at
once above the ore, and to save the silver which is thus
carried out, by allowing the washings to flow through a
series of launders, together with an additional stream
of water.   In the launders is placed a quantity of wood
shavings, or some similar material, on and among which
the silver chloride, together with base metal chlorides,
is deposited.   The shavings, with the metal chlorides,
are then gathered up, and the silver is extracted, either
by leaching, or by a smelting operation in a crucible.
The flow of water through the ore is continued, until
a test with calcium sulphide gives no precipitate.

140.   If the ore contains lead chloride, it is washed
with cold water until the greater part of the copper
and iron chlorides, and salt, are removed.   It is then
treated with hot water, as long as any metal can be
extracted, after which it may be necessary to again
apply cold water, to cool the ore before admitting the
hypo, so that too much base metal may not be extracted with the silver.   Hot water, if applied at first,
increases the solubility of silver chloride in the solution
of base chlorides and salt; but after these are in the
main removed, by means of cold water, it may be used
with advantage.   If, however, it were immediately followed by the hypo, more base metal would be extracted
than would be the case if the ore were cooled; for, as
before remarked, there are base metal compounds in

the roasted ore which are insoluble in both hot and cold water, but which are soluble in the hypo, especially if it is warm.

Lead chloride is almost insoluble in cold water, but dissolves readily by the aid of heat. The sulphate is not dissolved by water, hot or cold. Hence it is better that lead should be chloridized in the roasting, because it can then be removed by washing. If lead is extracted in the silver leaching, it cannot, like copper, be retained in the matte when melting, but inevitably goes into the bullion, because its sulphide is easily reduced by iron at a red heat, while the copper sulphide is not.

141. From this point the leaching and precipitation are carried on exactly as directed, after the gold leaching, for concentrations containing gold and silver. If the ore contains a little gold, a portion of it may be obtained with the silver, owing to the formation, during the roasting, of the peculiar gold chloride described (30).

142. *D.—Concentrations rich in Gold and Silver, and containing much Lead, etc.* It appears that certain rich ores, containing much lead, and other obstructive metals, do not yield the gold well when treated in the ordinary way. To meet this case Ottokar Hofman has devised and patented the following modification of the process.

The ore is subjected to a thorough chloridizing roasting, then washed with water, as described in the case of silver-bearing concentrations, leached for silver, and again washed to remove all hypo. It is then removed from the vat, dried sufficiently for chlorination, returned to the vat and chlorinated. The gold is then leached out, and if there still remains a considerable quantity of silver, the ore is again leached with hypo for its extraction. The results are said to be very satisfactory.

143. *E.—Unconcentrated Ore.* This material, containing gold or silver, or both gold and silver, is treated in the same way as described for concentrations of similar character, except that, if it is only moderately charged with sulphides, the furnace is made hotter before being charged, and the roasting is effected more speedily, and with less draft in the furnace, the rule being only to allow as much as suffices to prevent the escape of fumes at the doors.

If, as often happens, the ore contains too little sulphur to effect the chloridation of silver, it is necessary, if that metal is present, to add a certain quantity, say four or five per cent of crushed pyritous ore, one per cent of sulphur, or from two to three per cent of copperas. The pyrites or sulphur must be mixed with the ore before roasting, but copperas may be put in later, as described in certain cases occurring in the treatment of concentrations.

144. Unconcentrated ores are not often treated for gold by lixiviation. They are crushed as coarsely as is compatible with good roasting, but nevertheless are often troublesome to leach, on account of clay, talc, etc., so that, while concentrations can be leached in a bed of from two to four feet in thickness, without difficulty, or special appliances, unconcentrated ores will, in many cases, not admit of more than ten inches, and, in extreme cases, they cannot be worked by leaching.

This difficulty is sometimes overcome by crushing the ore in a wet battery. The water removes the slimy matters which impede filtration, while the heavier portion of the ore is retained in "catchpits." This is in fact a species of concentration, although a very imperfect one. As the slimes usually contain a considerable quantity of silver, they must also be preserved, and, since they cannot be leached, must be treated in some other way. Amalgamation is usually adopted. On the whole, the plan is scarcely to be recommended. In cases of difficult leaching the suction pipe, (54) may be used.

145. When about to begin leaching with the aid of a suction pipe, it must be filled with liquid. When operating on ore which is not chlorinated with gas, this is done by connecting the pipe with the water supply, and passing the water upward through it, and the ore mass, as directed in beginning the washing of silver ore (139). If a vent pipe (55) is in use, it must be plugged as soon as the space below the filter is full of water. If the ore has been chlorinated for gold, it is not desirable to introduce the water in this way, because the solution of gold would be too much diluted. It is then better to let the pipe down into the trough, and stop the end with a tight plug. On introducing water into the vat, as described in the gold leaching, it passes through the ore, and, displacing the air, fills the pipe. When the water stands permanently above the ore, and air bubbles have ceased to rise, the plug is removed, the vent pipe stopped, and the leaching is allowed to proceed, aided by the weight of the column of liquid in the suction pipe, or, more correctly, by the pressure

of a corresponding column of air.   The suction pipe may be filled before charging the vat with ore, if so desired, and in the case of silver ore the water may even rise to the top of the filter, thus avoiding all trouble from air; but when chlorine gas is to be used, the water must not cover the aperture through which the chlorine is admitted.

To reduce friction, and give the best effect, the the suction pipe should be large; say, for a ten foot pipe, one and a half or two inches in diameter.   It is impossible for it to take air at the lower end, with either of the arrangments mentioned, nor can any enter at the top as long as the vent pipe is closed and there is liquid in the vat.   I have observed that a fall of six feet more than doubles the flow of liquid through a bed of ore fifteen inches deep.

146.   *F.—Ores Containing Coarse Gold, or an Alloy of Gold and Silver.*—If the gold is too coarse, it will not be entirely dissolved in a single operation by the above described process of chlorination, or if it is alloyed with a considerable proportion of silver, unless in very fine particles, the   chlorination may be obstructed by the formation of a crust of silver chloride on each of the particles.   Gold which is too coarse is more easily col- lected by amalgamation; yet it can be dissolved by leaching with chlorine water, that is, water through which chlorine has been passed until no more is absorbed. An alloy of gold and silver can be leached out with brine of common salt, saturated with chlorine.   The chlorine acts on both of the metals, and the brine dis- solves the chlorides as fast as they are formed.   The

dissolved metals may be precipitated together, by means of plates of copper.

147.   The Mears process, recently introduced, seems to be well adapted to this class of ore.   It is conducted as follows: The roasted ore is treated in rotating, lead lined, iron cylinders, with the addition of water and chlorine, the latter being forced into the cylinders under heavy pressure.   The gold is said to be dissolved very rapidly, not more than two hours being required.   The pulp is then thrown into leaching vats, and leached in the usual manner.   An addition of salt to the charge in the cylinder would be useful in case of an alloy of gold and silver.   The advantages of applying the chlorine, or chlorinated brine, under pressure, have long been known.

148.   *Calcium Polysulphide.*—This precipitant for silver is made by boiling lime and sulphur together in water. Lime is calcium oxide, and, when boiled with sulphur, a part of it is decomposed; the calcium combines with sulphur, making calcium sulphide, while the oxygen, and some undecomposed lime, combines with another portion of sulphur, making calcium hyposulphite, which is the same as the solution used in leaching silver.

When the solution of sulphide is used for the purpose of precipitation, the water that it contains is added to the leaching solution, which would thus be gradually diluted but for the hypo which the precipitant also contains.   This is the reason why the volume of the leaching solution frequently increases, notwithstanding a certain amount of waste.   Not only is hypo formed in making the calcium sulphide, but the sulphide itself absorbs oxygen from the air, and changes into hypo to a certain extent.

Calcium sulphide may be made in an iron pot, over a fire, with frequent stirring. A neater way is to make it in a wooden tub, using a jet of steam for the heating. It can thus be made in the vat in which it is kept. The vat is two-thirds filled with water, and steam is admitted. As soon as the water is hot, freshly slaked lime is mixed with it, and flowers of sulphur added by rubbing through a sieve. When the sulphide is made in this way, two or three hours boiling suffices.

149. The proportions of lime and sulphur vary with the quality of the lime. About 1½ pounds of average lime is required for one pound of sulphur; 75 pounds of lime, 50 pounds of sulphur, and 120 gallons of water will produce a solution of suitable strength, and will suffice for the precipitation of from 25 to 50 pounds of silver, according to the quantity of base metal in the lixivium. If a direct fire is used for the boiling, the water which evaporates must be replaced, but if steam is used, a little less water than the prescribed quantity should be taken, as some will be added by the condensation of steam. It is however to be observed, that with an insufficient quantity of water, the operation proceeds very slowly. The proper strength of the solution is indicated by a density of about 10° Beaumé. If below 6° it will dilute the hypo too much, when used for the precipitation.

163. The chief points to be observed in making the sulphide are, that the lime must be slaked before use, and must not be in excess. If it is not previously slaked, time and material will be wasted. If used in

too great proportion, insoluble compounds are formed, or, at least, compounds which are not sufficiently soluble to be useful.   An excess of sulphur is not injurious. When the solution is properly made, with good lime, it is nearly of the color of strong coffee; it deposits, on cooling, few, if any, needle-shaped, yellow crystals, and the residue consists only of a small quantity of matter, of a dirty greenish color.

150.   *Calcium Hyposulphite.*—This solvent may be made by passing air, and the fumes from burning sulphur, or from sulphuric acid and charcoal heated in a retort, through a solution of calcium polysulphide until the latter is colorless.   It is better to buy a barrel or two of sodium hyposulphite in crystals, with which to make the leaching solution to begin with.   In use, with calcium sulphide as the precipitant, the sodium hypo soon disappears, being replaced by calcium hypo through the chemical reactions which take place.

151.   The strength of the solution to be used for leaching depends somewhat on the composition of the ore.   If this contains but little base metal, the solution may be quite strong, and is even used warm.   But, in general, a strong solution would extract too much base metal.   It may be made by dissolving two pounds of crystallized sodium hyposulphite in each cubic foot of water, or about 26½ pounds to 100 gallons. If it is then found to dissolve too much base metal, which may be ascertained by an examination, or an assay, of some of the precipitate, the strength is reduced by an addition of water.   After it has been used the density cannot be relied on, as it then contains

6

other substances besides hyposulphite. A good guide is the taste, which should be very sweet during the first stage of the leaching, if the ore contains much silver. The solvent power of the solution may at any time be tested, thus: Dissolve 5.25 grains of pure silver in nitric acid; precipitate as chloride, by adding a little hydrochloric acid. Wash the precipitate three or four times with water, to remove every trace of acid. One fluid-ounce of the leaching solution should dissolve the whole of the silver chloride. Some operators, when in want of leaching solution for silver ores, obtain it by treating with calcium polysulphide the solution of base metal chlorides resulting from the preliminary washing of the roasted ore. The metals are thrown down as sulphides, and the solution then contains calcium chloride, together with the calcium hyposulphite previously existing in the precipitant (148).

152. *Working Test.*—The apparatus used is represented in Figure 1, Plate 6. The generator and wash-bottle are made of two wide-necked bottles, jars, or flasks, with corks, and some glass tubes. The corks should be soaked in melted paraffin, and the S tube is enlarged at the upper end, so as to form a small funnel, into which to pour the acid. If preferred, a glass generator and two-necked wash-bottle can be bought of J. Caire, in San Francisco.

A chlorinating vat is made of a common wooden pail, or small tub, in the bottom of which, near the side, a hole is bored and a cork inserted. Through the cork is passed a piece of ¼-inch glass tube 4 inches long. The cork and tube must not project above the bottom of the pail inside. A wooden cover is made to fit into the pail half an inch below the rim, and in it is a hole, fitted with a cork and tube similar to those in the bottom. The pail, while dry, is thoroughly coated inside with melted paraffin, which is caused to soak into the wood a little by the aid of heat. A filter is made in the vat by means of a layer of pebbles, covered with a piece of moistened grain sack, or similar material.

A weighed quantity, from 10 to 20 pounds of the pulverized ore, or concentrations, is dead roasted or chloridized, in a small reverberatory furnace, with the precautions indicated, according to the character of the ore. When cooled, it is slightly moistened, and thrown on the filter in the vat. A space of not less than half an inch must be left between the surface of the ore and the cover, which is now placed, and luted with dough, or with a paste made from a mixture of equal parts of flour and paris plaster, which will not crack in drying.

Chlorine is generated from manganese and hydrochloric acid, the latter being more convenient for small operations than the sulphuric acid and salt used on the large scale. Three ounces of manganese are put into the generator, moistened with water, and warmed on a sand bath resting on a small coal-oil stove. The acid is gradually added by means of the S tube. The exit tube of the generator is connected with the wash-bottle, and that with the glass tube in the bottom of the vat, by rubber tubing.

When a glass rod, dipped in ammonia and held to the tube in the cover of the vat, causes the formation of dense fumes, indicating the escape of chlorine, the surplus gas is conveyed out of the room by a rubber tube connected with the tube in the cover.

The chlorine is allowed to pass through the ore for an hour, more acid being poured into the generator when required, to maintain the evolution of gas. The waste pipe is then closed by a pinchcock. The wash-bottle is disconnected from the generator, but not from the vat, unless it is required for another operation, in which case the glass tube in the bottom of the vat is effectually, and conveniently, closed by immersing its lower end in melted stearin, paràffin, or tallow, contained in a small cup (dry cup), which is then allowed to congeal.

After the lapse of from 20 to 40 hours, as may be required, the cover is removed. A rubber tube, connected with the glass tube in the bottom of the vat, is arranged to deliver into a glass, or porcelain vessel, capable of containing about a gallon, and is closed by a pinchcock.

Water is now sprinkled on the ore, and when the latter has settled, more water is poured in until it is covered. After half an hour, the pinchcock on the discharge pipe is adjusted so as to allow the leach to flow in a slow stream, water being poured upon the ore from time to time to keep it covered. The leaching is continued until a sample, received in a test tube, no longer gives the slightest precipitate on addition of solution of iron sulphate. A more delicate test is that with tin protochloride, which, with the slightest trace of gold, gives a purple coloration.

The gold is precipitated by adding to the leach a strong solution of iron sulphate, which is thoroughly mixed by stirring with a glass rod. To ascertain if enough of the iron sulphate to precipitate the whole of the gold has been used, a drop of the liquid is transferred, by means of the glass rod, to a porcelain dish, or a saucer, and is brought into contact with a drop of solution of potassium ferridcyanide, or "red prussiate of potash." If an intense blue coloration is not produced, more iron sulphate is required.

The gold requires 12 hours to settle, after which the greater part of the clear liquid may be removed by means of one of the rubber tubes, applied as a siphon. The remainder, with the gold, is thrown on a paper filter, or the whole of the liquid may be passed through the filter, to insure the collection of every particle of the metal.

The sides, and bottom of the vessel in which the precipitation was effected, and the glass rod with which the stirring was performed, are carefully wiped with pieces of filter paper, held in the forceps, to remove adhering gold, and the paper is added to the gold on the filter. The filter, with the metal, is dried in the funnel, and is then placed in an assay crucible, together with an ounce or more of litharge, and a little borax, and smelted. The filter reduces a sufficient quantity of litharge to metallic lead, for the collection of [the gold, which is then separated by cupellation, weighed, and the result compared with the assay of the ore.

If the ore contains silver, the gold on the filter may be washed with water, then, with the entire filter, drenched with ammonia, and again washed, before being dried. This will remove any silver chloride which may be present, or the bead may be inquartated and parted in the usual way. After the extraction of the gold, the ore, which in this case should have been roasted with salt, is leached with hypo for silver as long as a precipitate is produced by the addition of a drop of calcium sulphide to a sample of the leach.

The silver is then precipitated with calcium sulphide, as in the large way, except that, as it is not necessary to preserve the hypo, an excess of the sulphide is used, so that not a trace of the metal may be lost. The precipitate is coagulated by heating on a sandbath, separated from the liquid by filtration, dried, and dressed on the filter with litharge and borax, then fused in a crucible, with the addition of a little nitre to prevent the production of too much lead by the action of the sulphur on litharge. The lead button obtained is cupelled, and the resulting silver bead, after weighing, should be subjected to parting, as it may contain a little gold.

The tailings are dried and weighed. The loss of weight found to have been sustained by the ore in the working is reduced to percentage. An assay of the tailings is then made, and from the result the same percentage is deducted. The remainder is the loss per ton of ore by insolubility. This is added to the amount per ton extracted, and the sum deducted from the assay value of the original ore. The remainder is the loss per ton by volatilization, dusting, and other causes.

153. A smaller test may be made in the laboratory. Half an ounce, or an ounce, of ore, is roasted in the muffle, with the precautions indicated under "Assaying Concentrations" (160), with or without salt, as required, and is chlorinated in the apparatus represented in Fig. 2, Plate 6, formed of two glass funnels. In the

neck of the lower funnel some fragments of broken glass are placed, and on them a filter is constructed of rather coarsely powdered glass. Upon the filter the moistened ore is placed, the upper funnel is then arranged as a cover, and luted with a paste of flour and paris plaster. Two ore sample bottles, fitted with corks and tubes, as in Figure 1, suffice for a generator and wash bottle. The corks should be saturated with paraffin or tallow.

This operation has the advantage that, after the precious metal has been extracted and collected, as in the larger test, the whole of the ore and glass filter, dried in the funnel, can be dressed as an assay and smelted, so that it not only shows what can be extracted, but it also gives the absolute loss by insolubility, and, by difference, the loss in roasting, without complications arising from a change of weight in working.

154. *Change in Weight.* When working on the large scale by lixiviation, it must be remembered that the ore not only gains or loses weight in the roasting, but also loses in the leaching by the amount of soluble matter, of whatever kind, extracted. The tailings assay, therefore, does not represent the real loss by insolubility unless corrected. The change of weight sustained in roasting and leaching, is approximatively obtained by weighing, roasting, leaching, drying, and reweighing a number of small samples, using the same proportion of salt, and, as nearly as possible, the same heat in roasting, as in the large way. The loss of weight thus found, when reduced to percentage, gives the correction to be applied to the assay value of the tailings. Thus, if tailings assay 2.6 oz. per ton, and the loss of

weight in roasting and leaching has been found to average 22 per cent, the real loss per ton of crude ore, from insolubility, is 2.6—0.57=2.03 oz.

155. *Loss in Roasting.*—The only reliable method of determining the loss of gold or silver in the roasting of ore on the large scale is, to dry, weigh, sample, and roast a quantity, say from one to ten tons; then, after cooling, to weigh and sample again. It may weigh more or less than before, according to the character of the ore, the quantity of salt used, and the manner of roasting. The assay of the unroasted ore, coupled with the total weight taken, gives the quantity of precious metal in the lot before roasting, and the assay of the roasted ore, with its total weight, gives the quantity remaining after roasting. The difference is the loss, caused by dusting and volatilization. To make the test reliable, the furnace must be thoroughly cleaned, both before and after the roasting; if this is neglected, the weight obtained for the roasted ore will be incorrect.

As a part of the precious metal lost in the roasting may be recovered from the flues and dust chamber, while, besides the losses already discussed, others may occur in the leaching, collecting, and melting of the metal, it will be seen that a final result can only be reached by means of a general and complete clean-up; yet, as this cannot be had very frequently, the investigations described are indispensable as guides in working.

It is best, when any doubt is entertained as to the best way of roasting a given ore, to experiment in the laboratory on ounce or half ounce samples taken from crushed lots of considerable magnitude. The method employed by some metallurgists, of taking samples from the furnace at intervals during the progress of the roasting, and assaying them with the view of ascertaining the loss at successive stages, is open to the objection that it cannot be known exactly what correction must be made for the changes of weight sustained by the ore. It is, however, very proper to make such trials, because a heavy loss might be thus indicated in time to alter the treatment of the charge. The percentage of the silver which is soluble must be considered in connection with the loss by volatilization, in order to arrive at the most profitable manner of roasting; for it is better to make rich, rather than poor, tailings, if the difference should go up the smoke stack.

156. *Solubility Assay.*—This is commonly, but incorrectly, called a " chlorination assay." It is made as follows: From a quantity of the roasted ore, two assays are weighed out, one of which is leached on a filter with hypo as long as any metal can be detected in the filtrate; it is then washed and dried. The leached and the unleached samples are then separately smelted, and the resulting lead buttons cupelled and parted. The result obtained from the unleached sample gives the assay value of the roasted ore; that from the leached sample the portion of the silver which is insoluble in hypo.

If the average change of weight which the ore sustains in roasting has been ascertained, a corresponding correction, applied to these assays, will show the loss in roasting, if an assay has been made of the unroasted ore; thus:

Raw ore contains 56 oz. of silver per ton.

Loss of weight in roasting=5 per cent; therefore, Roasted ore should contain 56 + 2.9=58.9 oz. per ton, but actually contains   57.0 " " "
___
The loss in roasting is therefore   1.9 " " "
___

Unleached sample contains 57 oz. per ton
Leached   "   "   1.5 " " "
___
Soluble silver   55.5 " " "   =97.3 per

cent of the silver remaining in the ore after roasting. But a ton of roasted ore which has lost 5 per cent in weight corresponds to 1.052 ton of raw ore, containing 58.91 oz. of silver, of which we extract 55.5 oz. The total loss therefore is 3.41 oz., in 58.91 oz., or 5.79 per cent nearly, which, however, would be a better result than is often obtained.

PLAN OF WORKS.

157.   Plate 7 will give the reader a good idea of the general arrangement of a plant for treating, per day of 24 hours, from 1 ½ to 3 tons of concentrations containing gold and silver. The appended description will also serve as an inventory of the principal articles required :

A—Two hearth roasting furnace ; *A A'*—The same extended to three hearths ; *B*—Drier ; *C*—Smoke stack ; *D*—Floor, level with top of drier, with store-room beneath; *E*—Hopper; *F*—Boiler, for heating purposes, set in flue; *G*—Passage way under flue ; *H*—

Melting furnace; *k*—(dotted) Small roasting and drying furnace for precipitate—inside drier; *l*—Sheet-iron muffle for drying gold; *m*—Assay furnace; *n*—Doorway to laboratory; *o*—Steps to lower floor; *p*—Tank for hypo; *q q q q*—Leaching vats; *r r*—Silver precipitating tubs; *s*—Calcium sulphide vat; *t*—Well and pump; *u u u*—Filters for precipitate; *v v*—Gold precipitating tubs; *w*—Iron sulphate vat; *x*—Small tub to receive washings from *v v;* *y*—Filter tub for waste solution; *z*—Gasometer. or blower; a—Chlorine generator; b—Wash-bottle; c c—Cars for removing tailings; d d—Tram-ways, 6 feet above floor; e—Press for silver precipitate.

ASSAYING CONCENTRATIONS

158. It is well known that many experienced assayers, although generally reliable, do not obtain correct results from the assay of gold-bearing sulphides, very rich ores of silver, or cement copper containing gold, and contaminated by basic chlorides. A few facts and suggestions on this subject may not be out of place here. As this is not a treatise on assaying, the reader is supposed to be familiar with the apparatus and manipulations to be employed.

A common practice among assayers, and by some deemed indispensable, is to roast the assay of gold-bearing sulphides. It is better not to do so unless they contain much arsenic, and even in that case, scorification is preferable.

When an accurate assay of an ore which is new to the assayer is desired, it is best to make several assays by different methods. If the assayer knows how to properly clear, weigh, and part the bead, and to make the necessary allowance for silver in the litharge or lead employed, the presumption of correctness is in favor of the highest result, because, although it is easy to lose gold or silver in an assay, it is impossible to extract more than the ore contains. In important assays, such as those on which the purchase of a lot of ore is based, the carefully taken sample should be thoroughly dried at the heat of boiling water, passed through a seive of at least 80 meshes to the running inch, and then very carefully mixed.

It is generally assumed, and so implied in books on the subject, that if two assays, made at the same time, and in the same manner, give correspondiug results, they are correct. This is a great mistake. Two chemical operations conducted under precisely similar conditions should give similar results. Their agreement proves that no accidental loss has occurred, nor any error in weighing, but it does not prove that the method adopted is that by which the highest results can be obtained.

159. There are two principal methods of assaying ores for gold and silver—by crucible, and by scorification. The crucible assays may be divided into two classes; firstly, that in which a thoroughly oxidized slag is produced; secondly, that in which a matte containing the minimum proportion of sulphur is formed.

The scorified assay is always a completely oxidized assay, and is the most certain in its results, but is subject to certain inconveniences. Either a very small quantity of ore is operated on, or, a very large quantity of lead is obtained, and it usually requires more time than a crucible assay. Yet it is indispensable as a check, for in some cases it gives higher results than can be obtained from the crucible, especially with some arsenical compounds.

Each kind of assay is subject to modifications as to the quantity, kind, and respective proportions of fluxes used, depending on the constituents of the ore; whence it is difficult to give rules to meet the requirements of all cases.

Concentrated gold bearing sulphides, chiefly iron pyrites, with or without an admixture of other sulphides containing silver, are assayed by the different methods as follows :

160.   *1st Assay.*—Half an ounce of the finely pulverized and dried ore is roasted in a roasting dish, in the muffle.   To prevent loss by decrepitation, the assay is covered by another roasting dish inverted.   A small hole should be bored through the cover, and a few notches broken in its edge, so that the arrangement resembles a small reverberatory furnace.

The heat is kept very low, and after a time the dish is taken from the muffle and cooled a little, to prevent decrepitation, and consequent loss, on uncovering it. The cover is then removed, and the ore well stirred and turned over, again covered, and returned to the muffle.   This is repeated until it is found that the assay may be uncovered and stirred within the muffle, without danger of loss by decrepitation.   The heat is then increased to bright redness, and so maintained until the ore is odorless.   After cooling, the assay is dressed thus :

```
Ore (roasted)....................
Litharge......................... 2 to 4 ounces
Sodium bicarbonate...............½       "
Flour............................40 grains.
```

Mix, and place in a No. 8 French crucible, containing ½ ounce of borax in crystals, and cover with salt. Fuse quickly in a strong fire.   Allow the assay to remain five minutes in the fire after fusion, and pour.

The lead button should weigh about 300 grains.   If it varies much from that weight, the assay should be repeated, with a larger or smaller quantity of flour as the case may be.   A duplicate assay should be made, with either more or less litharge, as a check.   The lead buttons are cupelled as usual.

161.   *2d Assay:*—

Ore (not roasted)............	½	ounce
Litharge.......... ....2 to 4		"
Sodium bicarb ....½		"
Dried borax................... .½		"
Pulverized nitre................420 grains.		

Mix; put into a No. 8 French crucible, and cover with salt.   Heat the assay very slowly until fused; raise the heat to bright redness, and keep it so for five minutes at least; then pour.

The button should weigh about 300 grains, and be soft.   No matte should be formed.   If these conditions are not fulfilled the assay is imperfect, and must be repeated with a different proportion of nitre, more if the button is too large, or if a matte is formed; less if it is too small.   Theoretically, 25 parts of nitre convert 100 parts of lead into litharge; in practice the proportion varies with the purity of the nitre.

162.   *3d Assay.*—

Ore................................¼		ounce
Litharge.................... ...2 to 4		"
Sodium bicarb....................½		"
Dried borax.................... .....½		"
Pulverized nitre ....300 grains.		

Place the mixture in a No. 8 crucible; cover with salt; fuse very slowly.   When ebullition ceases, under a good red heat, add 20 grains of charcoal in lumps, a portion at a time.   When the evolution of gas produced by the charcoal ceases, cover, and heat strongly for ten minutes; then seize the crucible with the tongs and shake it, without removal from the fire. Again cover, and leave it in the fire for a few minutes longer, and pour.

In this assay, the quantity of nitre is sufficient for the complete oxidation of even pure iron pyrites. Any excess is decomposed by the heat, and after its decomposition the addition of 20 grains of carbon generally reduces a suitable quantity of lead, which, for 120 grains of ore, is about 150 grains. If it does not, the carbon alone is varied. The method is a little tedious, but gives very good results, especially as to gold.

163. In all the preceding assays the crucible is liable to extensive corrosion, unless the ore contains a considerable quantity of quartz. This may be prevented by the addition of half or a whole teaspoonful of pulverized quartz, which must of course be free from precious metal. Glass will answer the purpose, but rather more of it is required.

164. An assay which is dressed with undried borax is less liable to boil over if put at once into an intense fire, than if heated by slow degrees. When nitre is used rapid heating is not admissible, hence the borax is dried before use. It is well to dry it in all cases. Boiling over sometimes occurs in consequence of an evolution of carbonic acid gas (carbon dioxide) produced by the fusion of the alkaline carbonate, used as a flux, with silica. A large proportion of litharge, and less of the carbonate, will correct this, while forming an equally good slag. Whenever the assay tends to boil over, it may be cooled down by throwing a little dry salt into it.

7

165. *4th Assay.*—

    Ore . . . . 120 grains.

    Litharge 6000   "

Cover with salt; fuse quickly, and pour as soon as the assay becomes quiet. The lead button should weigh about two ounces, if the ore consists chiefly of iron pyrites. It must be reduced by scorification to a suitable size for the cupel.

166. *5th Assay.*—

    Ore . . . . ½ ounce.

    Sodium bicarb . . . . 720 grains.

    Litharge 300   "

A little flour and pulverized glass will do no harm, and in some cases are beneficial. Place in a No. 7 or 8 crucible, with a very little borax and three large nails; cover with salt. Fuse slowly; then give a white heat for half an hour or more. If, on seizing one of the nails with the tongs, rinsing it in the fluid slag, and tapping it against the side of the pot, it is found to be free from adhering lead globules, the other nails may also be removed; but if the nail shows adhering lead, which cannot be washed off, more time is required.

The matte and slag should retain no lead in any form; hence the weight of the button should correspond with that of the litharge used, or may be in excess, if the ore contained lead. It must be soft, and separate well from the matte. If it does not, more soda is probably required. 300 grains of pure litharge contain 278.5 grains of lead, but five per cent may be allowed for volatilization.

This is a very convenient assay, requiring but little modification for the different sulphuretted ores. It is not liable to boil over, nor is it so destructive to the crucible as are the other methods described. If properly made, it gives as high results as can be obtained by any other method, in most cases, both as to gold and silver.

Arsenical mattes cannot be assayed successfully by this method, and, in one instance of gold and silver bearing sulphides, though it gave the gold correctly, it did not yield as much of the silver as the assay by scorification.

167.   Oxidized coppery ores, and especially cement copper contaminated with basic chlorides, may be conveniently assayed for gold and silver by a similar process, as follows:

> Ore or cement copper..............½ ounce.
> Sodium bicarb.................720 grains.
> Litharge......................300   "

Some flour, or charcoal, borax, and enough sulphur to convert all the copper, the sodium, and the lead from the litharge, into sulphides. An excess of sulphur does no harm. The mixture, covered as usual with salt, is first fused; the nails are added, and a strong heat is continued for half an hour. The copper remains in the state of matte; the nails reduce the lead sulphide, and take up all excess of sulphur. Massive copper, and other metals, may be treated in the same way, if first broken, or cut into fragments.

168.   *6th Assay.—Scorification.—*

> Ore .... 60 grains.
> Granulated lead...............480   "

Mix in a scorifier, and cover with another ounce of granulated lead. Place the scorifier in the muffle, under a low heat, and increase the temperature gradually, until the "bull's eye" appears—that is, until the surface of the melted lead is clear in the centre, and the fused, or partly fused, ore forms a ring around it. Keep up a cupelling heat, until the slag quite covers the lead, then, with the tongs, add some lumps of borax, and give a strong heat. Stir with a red hot piece of stout iron wire, to detach pasty lumps from the side of the scorifier. Pour when the slag has become so fluid as to run completely off the stirrer. If the lead button is too large, return it to the scorifier, after detaching the cooled slag, and oxidize it until of the proper weight.

169. Nearly clean silver glance is conveniently assayed by mixing it with two or three parts of litharge, melting quickly, and pouring as soon as well fused. The assay No. 5 is well adapted to heavily sulphuretted silver ores, containing antimony, arsenic, and zinc, as well as copper. It is rarely necessary to roast silver ore for assaying.

170. In assaying rich silver ores, that part of the cupel which is saturated with the lead oxide must be pulverized, dressed with litharge and flour, smelted, and the resulting lead cupelled, as a considerable quantity of silver is thus added to the result. This is quite important.

171. Tailings of concentrated and roasted gold-bearing sulphides are treated precisely as No. 1, omitting the roasting. The reason why as much as 40 grains of flour must be used in an assay of 240 grains of the roasted ore is, that the iron peroxide which it contains consumes the carbonaceous matter, being reduced to protoxide, and if only the usual quantity of the reducing agent were used, little or no lead would be produced.

ADDENDA.

FLUE COVERING.

172. A very convenient covering for horizontal flues, which can be removed and replaced easily, is shown in the annexed diagram. It consists of a kind of 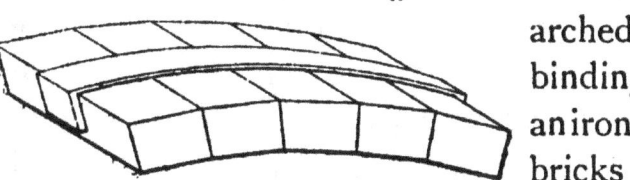 arched tile, made by binding together, with an iron clamp, as many bricks as may be required by the width of the flue, each tile, or section, having the length of one brick. The flue is covered with these sections, and the interstices are filled with mortar, or a mixture of clay and sand.

FILTERS.

173. Filters for dumping vats are made with a bed of twigs, in place of gravel, and over the twigs is a piece of cloth, which is held close to the inner surface of the staves by a wooden hoop. The objection to burlap or other cloth is, that it is speedily destroyed by chlorine gas. Attempts have been made to preserve it, by soaking it in coal tar and asphalt, but without much success. It occurred to me to use asbestos cloth, but I found none that was not too closely woven. Clay tiles, of the kind known as "biscuit ware," have been suggested for filters, and it may be that basket work, made of rattan, would answer, especially if soaked in petroleum, or ozokerite.

COPPER.

174. When concentrations which are to be chlorinated for gold, contain also copper, the roasting is not pushed to the decomposition of copper sulphate. The latter is converted into chloride by an addition of salt, a short time before discharging the ore from the furnace. The leach is then green, and, after precipitating and settling the gold, it is drawn off, either through a filter or not, as desired, and led into tanks containing iron, where it is allowed to stand until the tanks are again required. The iron goes into solution, and the copper is precipitated in the form known as "cement copper," which is collected periodically, washed on a filter, with hot acidulated water, dried and sold. As it may contain gold it should be assayed for that metal.

MEANS OF DRAWING LIQUID FROM PRECIPITATING VATS.

175. Some operators use wooden faucets for this purpose. One objection to faucets is, that a portion of the precipitated gold lodges in the bore, and it is necessary, on opening the faucet, to place a bucket under it to receive the first liquid which flows. Faucets often give trouble by leaking, especially when used for metallic solutions, which have a tendency to cause wood to shrink. Plugs are also objectionable on account of their liability to leak. If used they must fit flush with the inside of the tub.

A piece of rubber hose, applied as a syphon, answers very well, if prevented from drawing too near the bottom of the vat, by fastening a strip of wood to it so that the wood projects beyond the end of the hose. A better plan is, to insert in each end of the hose a tightly fitting piece of lead pipe, which is coiled once round in a spiral form. The hose having been filled with water, is laid across the side of the vat, the inner end dipping beneath the surface of the liquid.

Air cannot pass the curves in the leaden portion of the syphon while it is full of liquid, so that if the workman neglects to lower it as the liquid in the vat subsides, no inconvenience results, and the flow is resumed as soon as the inner end is lowered by drawing the hose over the side of the vat.

RECOVERY OF ABSORBED GOLD.

176. The false bottoms, of perforated wood, and the chlorinating vats, should be burned when no longer required, as they contain a considerable quantity of gold. At my works at Melrose the ashes of the four false bottoms yielded about $90 worth of gold and silver. The vats were assayed by boring a hole in one, burning the chips and smelting the ashes, which gave only $6 per tub, and, as they were still serviceable, it would not have been profitable to burn them. I attribute the fact of the vats having absorbed so much less gold than the false bottoms, to the circumstance that they had previously been used in the Hunt, Douglass, and Stewart copper silver process, and were quite saturated with brine, etc., while the false bottoms were of new lumber. Both were tarred.

GENERATORS.

177. There are several forms of generator. That described differs slightly from those in general use. In Deetken's generator, with water joint cover, the wall of the water chamber is not high enough to allow of filling the vessel with water, to expel the chlorine. Kustel describes one in which the stem of the stirrer passes through a stuffing box instead of a water joint, and in place of a water joint for the main cover, is a shallow groove containing clay and a little water. I have used this joint, and it answers very well, but hardly allows of filling the generator with water. A stirrer is not indispensable; at Crosby's works in Nevada City it was not used.

I think the apparatus shown in Plate 5 has some advantages over the others. Generators for chlorine are sometimes made like a "carbonic acid gas" generator, so that any required pressure can be had, but much pressure is not required in the ordinary way of working ores with chlorine.

WASH-BOTTLES.

178. The wash-bottle described (68) is of my own contrivance. Kustel describes one, made from an acid-carboy, arranged like that which I have described for the working test. Deetken used simply the bell glass, standing in an open basin of water, with leaden pipes for the introduction and eduction of the gas, passing under the edge of the bell. This arrangement would seem liable to upsetting. I used one in which

the bell was secured vertically, in a tub of water, by means of a cross board fixed in the tub, with a hole to fit the bell. The gas was introduced by a leaden pipe, passing under the bell, and escaped through the neck of the latter, with which a rubber pipe was connected.

The objection to a bell standing in an open vessel containing water is, that the water becomes saturated with chlorine, which is constantly evolved from the exposed surface. This, though not important in a well ventilated room, may produce inconvenience in closer quarters. The carboy is less convenient than the barrel for the renewal of the water, besides being liable to accidental breakage. So, on the whole, I think that my contrivance is to be preferred. The barrel will be the better for being tarred inside; better still if coated with paraffin.

If there is any difficulty in finding a bottle which will admit the chlorine pipe through the neck, it will answer to pass the pipe through the head of the keg. The gas will find its way to the bell, and, if the keg is full of water, the bubbling of the chlorine will be as visible as if it were delivered immediately under the bell, which, however, must be set higher than in the figure in this case, unless the pipe be of lead, and curved so as to deliver under it.

It is not essential that the gas should be *seen* to pass, because it can always be *heard* bubbling through the water, so that a glass bell is not absolutely required, but it is convenient, enabling one to see that the proper quantity of water is in the keg.

The water should be renewed once in two or three

weeks, and that which is drawn out can be utilized in making the precipitant for gold from scrap iron.

SIFTING ORE.

179. The usual method of sifting the damp ore into the vats, by means of a shaking sieve worked by hand, is tedious. The roasted ore might be sifted dry through a rotatory cylinder sieve, by hand or by power, and afterwards moistened as usual; then thrown lightly into the vat, by being shaken off a shovel, and dis-- tributed by raking. This method would admit of the use of a fine sieve.

COST OF ACID.

180. Sulphuric acid of 60° Beaumé, or 1.76 sp. gr., containing, according to Ure, 69.31 per cent of "dry acid," costs four cents per pound in San Francisco. One pound of this acid is equal to 1.06 pounds of acid of 63° Beaumé, or 1.71 sp. gr., containing 65.25 per cent of dry acid, and costing 2½ cents per pound. From this it is evident that the weaker acid is the cheaper to use, unless freight costs 15 cents per pound. (Double freight is charged on acid by the Railroad Companies.)   For example:—

```
1 lb. of acid of 66° ........ 4 cents.
1 lb. freight .............. 15
                           ——————  19 cents.
1.06 lbs. of acid of 63°... 2.65 cents.
1.06 lbs. freight at 15 cts..15.9
                           ——————  18.55 "
```

Taking into consideration the freight on the carboys, which must be paid both ways if they are returned, though not equally, an addition of about 30 per cent must be made to the weight on which double freight must be paid, and about 20 per cent of ordinary freight, or 10 per cent of the first addition, making the cost of the acids respectively:

Strong acid per ℔. as above..19.   cents.
Freight on carboy going......4.5   "
    "   "   "   returning....0.45
              —— 23.95 cents.

Weak acid as above, 1.06 ℔s...18.55 cents.
Freight on carboy going..... 4.77 "
    "   "   "   returning... 0.47
              —— 23.79 cents.

The carboys are charged for at the rate of \$2.25 each, and if not returned their cost must be added to that of the acid. This makes an inappreciable difference in the relative cost of the two grades. Acid can be bought and freighted more cheaply if a carload is taken at once.

———

SALT, MANGANESE, AND ACID.

181. According to the chemical view of the production of chlorine from salt and manganese, under the action of sulphuric acid, 26.8 pounds of pure salt is sufficient for 20 pounds of pure manganese binoxide. But, in this connection, two points must be considered. Firstly, the commercial salt and manganese are never pure, the former not often containing more than 90 per cent of sodium chloride, while the latter is consid-

ered to be a good article if it contains 70 per cent binoxide. Secondly, the theoretical reaction is not perfectly realized in practice.

If the proportion of salt is too small, as compared with that of the manganese and acid used, the latter two will react, giving off oxygen, and forming manganese sulphate, so that the subsequent addition of more salt will not repair the error. A mistake in the direction of an excess of salt can be rectified by an addition of acid and manganese, but as salt is the cheapest of the three substances, this is less important. In the directions given (94) for making chlorine, the proportions are those in common use, and the salt is intended to be somewhat in excess of the other substances.

In case the quality of the salt or manganese should differ much from the assumed average, there might be a considerable waste of one or the other, whence it seems desirable that the reader should be informed as to the manner in which these substances may be assayed, in order that they may be proportioned correctly.

182. *Assay of Salt.*—Take a fair sample of the salt to be assayed. It must not be dried, because what is required is its value for the production of chlorine in the condition in which it is to be weighed for working, but, if too coarse, it should be crushed. Mix thoroughly. Dissolve 10 grains in about 50 grains, or rather less than ⅛ fluid-ounce, of distilled or rain water, or at least water which gives no precipitate on addition of a drop of solution of silver in nitric acid. Filter the solution through a small wetted filter of fine bibulous

paper. The filter must be well washed with water to completely remove the salt solution, and the filtrate and washings received in a flask.

To the filtered solution in the flask add, drop by drop, a solution of silver nitrate until no further precipitation of silver chloride takes place. Warm slightly, and shake vigorously, stopping the mouth of the flask with a greased cork, until the precipitate is found to settle readily, leaving the liquid clear. Now transfer to a small filter, the weight of which has been previously ascertained after drying on a water bath until it no longer lost weight. Wash the precipitate and the filter with hot water until the washings are tasteless. Dry the filter, containing the precipitate, on a water bath until two successive weighings give the same result. From the weight obtained deduct that of the filter; the remainder will be the weight of silver chloride produced by 10 grains of the salt.

If the salt were pure and dry, 10 grains would produce 24.53 grains of silver chloride, but if we say 24.5 grains it will be sufficiently accurate. Multiply the weight of silver chloride found by 100, and divide by 24.5; the result will be the percentage of pure salt in the sample. EXAMPLE:—The silver chloride obtained from 10 grains of the sample weighs net 19.3 grains; then the sample contains $1930 \div 24.5 = 78.77$ per cent of pure salt.

The silver chloride should be dried without much exposure to light. For a water bath, the cover of a vessel in which some water is boiling answers very well. The pulp scales used in assaying ore are sufficiently accurate for this work, if sensitive to $\frac{1}{10}$ grain.

183. The salt sometimes contains a considerable quantity of sodium carbonate, which neutralizes an equivalent quantity of acid. In the interior of the country, where this is most liable to occur, acid is very expensive on account of the cost of freight. Such salt should therefore be purified, by dissolving it in water, and exposing the solution to evaporation in shallow vats, when the purified salt will be deposited.

184. *Assay of Manganese.*—Take a short necked, flat bottomed flask of three ounces capacity, and fit into it a cork through which pass two glass tubes of about ⅛ inch bore. One of the tubes must extend nearly to the bottom of the flask; the other, about three inches in length, is bent to a right angle about an inch from one end, the shorter limb being passed through the cork. Now take the barrel of a small glass syringe, such as is used for medical purposes, the orifice of which must be enlarged by cutting off a portion of the beak. Fit a cork to the larger end, and through it pass the disengaged end of the bent tube, the other end of which is in communication with the flask. Place in the syringe some pieces of pumice stone, and moisten them with sulphuric acid; attach to the bent tube by means of the cork. A test tube suspended from the neck of the flask completes the apparatus, which is represented by the accompanying cut.

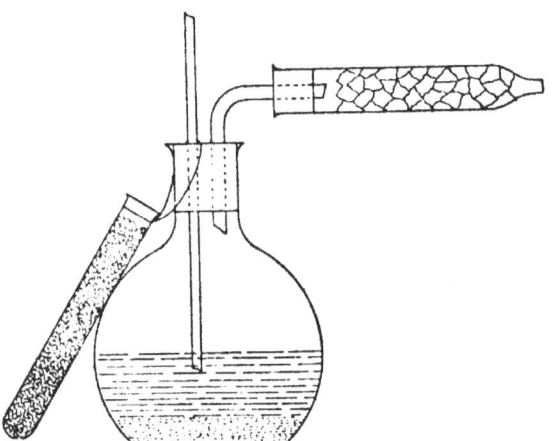

Weigh 99 grains of the finely pulverized manganese which place in the flask. Add one ounce of cold water, and then 150 to 160 grains of sulphuric acid. Replace the cork in the neck, with the glass tubes, and, applying the lips to the orifice of the syringe, suck air through the flask. This is done for the purpose of removing any gas which the acid may cause to be evolved from impurities in the manganese. Now weigh 150 grains of oxalic acid which place in the test tube, put the entire arrangement upon the pan of a balance which will indicate $\frac{1}{10}$ grain, and counterpoise it.

Remove the apparatus from the balance, throw the oxalic acid from the test tube into the flask, cork instantly and set aside for a short time. When effervescence ceases, warm slightly, which will renew it, and when no more gas is evolved draw air through the flask as before. Place the entire apparatus again on the scales. It will be found to be lighter than before, and the number of grains required to restore the equilibrium will be the percentage of binoxide in the sample.

This method, while not rigidly accurate, is sufficiently so for practical purposes. It depends on the circumstance that oxalic acid, in presence of sulphuric acid and manganese binoxide, is converted into water and carbonic acid gas, (carbon dioxide). The water remains in the apparatus. The gas escapes through the tube containing pumice imbued with sulphuric acid, which retains any vapor of water which might otherwise pass off, the last portion of the gas being drawn out by the mouth as directed. Thus nothing but the carbonic gas escapes, and as 99 grains of pure binoxide produce almost exactly 100 grains of the gas, it follows that the weight lost corresponds to the percentage of binoxide in the sample.

185. Having assayed the salt and manganese, the relative proportion in which they should be used remains to be considered. It has already been intimated that practice does not conform strictly to theory, not because the theory is incorrect, but because the conditions attainable in working do not admit of its complete realization. Hence it is found necessary to use sulphuric acid in excess of the chemical equivalent of the manganese, and as sulphuric acid is only useful in conjunction with a sufficient quantity of salt, the latter must also be in excess of the manganese.

For this reason, instead of only 27 pounds of pure salt to 20 pounds of manganese binoxide, we use 36 pounds; these quantities usually suffice for the chlorination of three tons of roasted ore. In order to ascertain how many pounds of the impure materials must be used, it is only necessary to divide the quantity of the pure substance required by the fraction representing the percentage found in the assay.

Example: Suppose the assay of the salt gave 87 per cent, it is required to know how much of it is equal to 36 pounds of pure salt, then $36 \div .87 = 41.37$ pounds; again, if the manganese is $66\frac{1}{2}$ per cent, then $20 \div .665 = 30$ pounds of the manganese to 41 pounds of the salt, which is about the proportion generally used, but either or both of the substances may vary from this supposition.

As to the acid, the quantity required depends so much on the temperature of the generator, and the nature and quantity of the impurities in the other materials, that no other rule can be given than this; it must be added in small portions as long as it continues to develope a useful quantity of chlorine. It will soon be discovered how much is required for a given quantity of salt and manganese.

The strength of the acid is indicated by its density, as measured by a Beaumé hydrometer, which is simply a floating gauge—the higher it floats the stronger is the acid, and the reverse.

SURPLUS CHLORINE.

186. The gasometer, suggested for the removal of surplus chlorine (98), might be made by attaching a leaden curtain to a disc of wood to which the suspending chains would then be fastened. A light tub, with wooden hoops, thoroughly soaked with paraffin or ozokerite, would probably answer the purpose.

The chlorine withdrawn from the vats, would, most likely, be contaminated with air, and as its re-use is not of great importance, the apparatus might be simply used to blow air through the ore mass, from below the filter, thus expelling the chlorine through a hose connecting with the hole in the cover of the vat, and conducting out of the room.

In this way there would be no difficulty in the construction of the bell, as either wood or iron could be used. Other methods of blowing air through the ore will readily suggest themselves.

Precipitating vats for gold are sometimes made in the form of rectangular tanks of wood, lined with sheet

7

lead. This vat, though more expensive than a wooden tub, has some advantages. There is no risk of absorption of the gold solution, and it is not necessary to add the precipitant until the vat has received its quota of solution. Then, if the solution contains much free chlorine, a jet of steam may be introduced and the chlorine driven off, after which the precipitation of the gold by means of iron sulphate proceeds better, and with a smaller consumption of the precipitant.

As it is necessary to heat the solution in order to drive off the chlorine in this way, it cannot be thus done so thoroughly in a tarred tub, on account of the softening of the tar. It could, however, be done by blowing air through the solution.

PRECIPITATING GOLD.

187. It occurred twice in my works that on collecting the precipitated gold it was found to be brick red, instead of having the usual brown color. It was also deficient in quantity. On the first occasion I could not account for this, but on the second I satisfied myself that the workman had not thoroughly mixed the precipitant with the solution, and being consequently deceived by a test, he had not used iron sulphate in sufficient quantity.

188. If the solution of gold contains lead chloride, the iron sulphate causes a precipitation of lead sulphate, which renders the subsequent washing of the gold difficult. This may be prevented by precipitating the lead as sulphate by means of sulphuric acid, or any soluble sulphate by which the gold will not be affected, as so-

dium sulphate. After settling, the liquid must be transferred to another vat, in which the precipitation of the gold is effected as usual.

In some cases the leach contains substances the character of which has not been determined, and which affect the purity of the precipitated gold. If the lixivium is allowed to stand several hours, a deposit is formed on the sides of the vat. The purified solution is then drawn off into another vessel, and the solution of iron sulphate is added.

SUSPENDED GOLD.

189. By taking some of the waste liquor from the gold tub, filtering twice through Sweedish paper, and then smelting the paper with litharge, I found that after 24 hours settling there remained gold in suspension, equal to $1.00 in value for each ton of ore leached. After 48 hours I still found half that quantity. The filtrates from these tests were boiled with zinc until all precipitable metals were thrown down. The precipitate contained no gold, which shows that iron perchloride does not dissolve gold in presence of the iron proto-salts, thus setting at rest a doubt which had arisen, as to whether or not the precipitated gold might be, to some extent, re-dissolved by remaining for many hours in contact with the solution of iron persalts, which is produced when gold terchloride is decomposed by iron proto-sulphate.

PRECIPITANTS FOR GOLD.

190. When the ore contains copper, the waste so-

lution from the tanks in which the copper is precipitated can be used for refilling the vat containing the precipitant for gold, as it contains a large quantity of iron protosalts, chloride and sulphate. The strength must be reinforced with copperas, or acid and iron. The waste liquor from the chlorine generator, consisting partly of manganese proto-sulphate, is also a good precipitant for gold. In order to utilize as much as possible of the chlorine which it contains, as well as any free acid which may be present, it should be placed in a covered tub, or small vat, with some scrap iron. In a day or two the chlorine and the acid will be saturated with iron, thus forming an additional quantity of precipitant.

Other preciptants which yield the gold in the metallic state are, oxalic acid, sulphurous acid, the antimony, arsenic, and copper lower chlorides, and animal charcoal. The latter is used as a filter on which the gold solution is poured, when the particles of carbon become covered with a film of gold. It is patented. In some European works the gold is, or was, thrown down as a sulphide, by means of hydrogen sulphide. This method has the disadvantage of precipitating also copper, and some other metals, if present.

SAND IN GOLD MELTING.

191. Many years ago I was employed as melter in a bullion assaying establishment, and was often called on to refine gold, which was done by granulating and re-fusing with nitre and borax, in a sand pot. It often

happened that the crucible broke in the furnace, causing a great deal of trouble.

Knowing that the pots were attacked not only by the metal oxides produced in the refining, but also by the potassa resulting from the decomposition of the nitre, it occurred to me to mix some clean quartz sand with the nitre, which answered so well that I never afterwards had a pot broken in the furnace.

DRYING CHAMBER.

192. For drying the gold, I suggest that a convenient arrangement would be a sheet-iron closet, in the form of a muffle, in the dust chamber, with a door which could be locked. In this the dishes containing the gold could be left with safety and convenience as long as necessary. A similar but larger closet could be used for drying the silver precipitate. Such a closet is represented at *l*, Plate 7.

VOLATILIZATION OF GOLD.

193. It has been stated by some authors that there is never much loss of gold in roasting ores, unless the roasting is performed too rapidly, and that the addition of salt makes no difference in this respect. I differ with them on this point, even when the statement is supported by so formidable a name as that of Plattner. But I think that illustrious chemist would yield to the logic of a pecuniary loss of three thousand dollars, especially when backed by other facts.

When I first entered on the business of gold chlorination on the large scale, I had a class of concentrations to treat which, while consisting chiefly of iron pyrites, and presenting no visible peculiarity, had nevertheless baffled many old operators, a fact of which I was not at the time aware. The ore was roasted in a three-hearth reverberatory furnace, with the addition of from 1 to 2 per cent of salt, on account of the presence in it of a considerable quantity of silver.

I was surprised to find that although the assays of the tailings were satisfactory, the gold, when collected, fell alarmingly short of the results which I had guaranteed, and I was of course obliged to make up the deficit.

Relying on the statement referred to, which I found in the only handbook on the subject within my reach, and being then a novice in this branch of metallurgy, I did not dream that a serious loss was taking place in the roasting furnace, especially as I had an expert metallurgist in reduced circumstances employed on the roasting, who had no more suspicion of the truth than I had.

Finding that I was sustaining some loss by inadequate tub room for the gold solution, and having consequently to draw the liquid from the gold-tub too soon after precipitation, I thought that I should find the whole loss to be in the leaching and precipitating department, and each time that an improvement was made, expected better results, so that I was led on from trial to trial, until the total loss reached the sum named.

I am not quite certain that I should ever have discovered the truth, but for the following accident. One day, I so far checked the draft of the furnace as to cause some fumes to come through the airholes and working doors, and a yellow sublimate on the masonry attracted my attention. On examination I found the sublimate to be very rich in gold, although there was none to be seen in it by the most careful washing. It also contained iron perchloride, and copper chloride, with some lead and other substances. This aroused my suspicion, and I at once did what I ought to have done sooner.

Knowing that the material could be roasted with little or no loss, if no salt was used, because some assays had been made in that way, I weighed two half ounces of a sample, and roasted them in the muffle side by side, under precisely the same conditions, except that to one of them I added 4 per cent of salt. · The roasting was purposely pushed to an extreme as to heat and time, and when the two tests were assayed, under exactly similar conditions, that which was salted was found to contain less than half as much gold as the unsalted one.

I then took some light fluffy sublimate from the flue of the roasting furnace, an assay of which gave me a value of some $600 per ton, chiefly gold. The quantity of this material was, however, very small, and the bulk of the matter in the dust chamber was not much richer than the average of the ore treated, a circumstance which indicates that the gold was actually to a great extent volatilized in some not easily condensable form. I also found that ·the ore sustained a loss of weight in roasting, equal to about 18 per cent, consequently the roasted ore ought to have been more than 18 per cent richer than before roasting, which was not the case.

If this is not considered to be sufficient proof that gold may be volatilized in the roasting of some ores with salt, the deficiency is supplied by the fact that, as soon as I made the necessary change by reserving the salt until the nearly dead roasting of the ore was finished, not only did the roasted ore assay 20 per cent richer than when raw, but the yield overran my guarantee, while the tailings nevertheless contained considerably more gold than before.

The moral of this is, never to neglect any precaution in the way of tests and assays, nor to place implicit reliance on rules laid down, or inferences drawn by others who have worked under different conditions.

I afterwards found that a very small quantity of salt, not more than three pounds to the ton, might be mixed with the crude ore without detriment to the gold, and with decided advantage to the extraction of the silver.

—

SOLUTION OF SILVER IN HYPO.

194. The statement that silver chloride dissolves in a solution of calcium hyposulphite is not strictly accurate. It is decomposed, exchanging constituents with the calcium salt, forming calcium chloride and silver hyposulphite. The latter combines with another portion of calcium hyposulphite, and forms a double salt, silver calcium hyposulphite, which is very soluble in water.

If calcium pentasulphide is added to such a solution in equivalent proportion, one-fifth of its sulphur combines with silver to form silver sulphide, which is pre-cipitated, together with the remaining four-fifths of the sulphur. The oxygen, and so-called hyposulphurous acid, which were combined with the silver, combine with the calcium of the decomposed precipitant. Thus the portion of the solvent which was decomposed by silver chloride is reproduced.

From this it follows that, for every atom of silver extracted from the ore, and precipitated by polysul-phide, the solvent gains a molecule of calcium chloride, without however having ultimately lost any hyposul-phite, and this is the reason why the density of the so-lution ceases to be an index of its solvent power. So far there is neither a loss nor a gain of calcium hypo-sulphite in the dissolving solution, but, as stated else-where, the precipitant always contains a quantity of that salt in watery solution, which being added to the solvent, increases its volume without impairing its strength, unless the precipitant is too much diluted.

In practice the increase or diminution in strength and volume of the solvent depends on the care exercised in saving as much of it, with as little addition of wash water, as possible, and in having the precipitant suffi-ciently concentrated. Kustel gives 6° Beaumé, as a minimum density for the latter.

WASTE OF SULPHUR.

* 195. The quantity of sulphur required for the pre-cipitation of silver is really only as 16 to 108, but a great waste occurs in the process described, from two causes: Firstly, there is always a certain quantity of

base metal dissolved with the silver, which also takes its portion of sulphur from the calcium sulphide; secondly, a large proportion of the sulphur is thrown down in a free state, and, in the usual course of procedure, is totally wasted by being burned off in the roasting of the precipitate.

The reason for this precipitation of free sulphur is that the precipitant is necessarily a pentasulphide, for the calcium mono sulphide is insoluble, and the bisulphide is only soluble with heat (*vide Regnault*), so that for our purpose we are restricted to the pentasulphide.

But in the precipitation, only one-fifth part of the sulphur of the pentasulphide can enter into combination with the silver, forming silver sulphide, and as the calcium combines with the oxygen and hyposulphurous acid which were combined with the silver, the remaining four-fifths of the sulphur are unoccupied and useless.

Base metals, when present in the solution, act similarly, combining with only a portion of the sulphur of the pentasulphide, only arsenic and antimony forming pentasulphides. Gold forms tersulphide, but its quantity is usually insignificant in the lixivium.

RECOVERY OF SULPHUR.

196. There are two methods by which the free sulphur may be recovered from the precipitate. The first, introduced, it is believed, by Ottokar Hofman, consists in subjecting the mass to heat, in a retort, and condensing the sublimed sulphur. The second, original with myself, and which seems preferable, is as follows :

The unwashed precipitate is placed in a vat with water, boiled by means of a jet of steam, and slaked lime carefully added.

The free sulphur combines with the lime, in the manner already described, producing at once the precipitating solution of calcium polysulphide, and much more quickly than when the fresh sulphur is used. The boiling can be done in a filter vat, and when finished the clear solution can be drawn off. The precipitate is washed by passing hot water through it, dried and roasted as usual. Three-fourths or more of the sulphur is thus recovered for re-use.

———

MATTE FROM THE SILVER.

197. The matte obtained in melting the silver precipitate is usually quite rich, and it seems to be impossible to completely desilver it by means of iron alone, at least in the presence of copper. The bullion is about .800 fine, even when there is a good deal of copper in the precipitate, because very little copper is reduced from the sulphide by iron at a white heat. It is, however, necessary that enough sulphur be left, in the roasting, to convert the copper into sulphide, for if too much oxide is formed a part of it may be reduced by the blacklead pot, metallic iron, or by charcoal if present, and thus contaminate the silver. Antimony and arsenic, though their sulphides are reduced by iron, yet volatilize, or combine with a further portion of iron and remain in the matte. Lead, if present in the precipitate, cannot be kept out of the bullion. Gold, when melted in the presence of sulphur and iron, forms a black, brittle mass,

containing much iron, but the gold sulphide obtained with the silver precipitate is mostly reduced in the melting, and the metal is found alloyed with the silver.

———

MELTING FURNACE.

198. I have seen, in the mining regions, many barbarously constructed melting furnaces, some of which were very inconvenient to work with. A description of the furnace which I build for silver melting, where economy is an object, and of which Figures 1 and 2, Plate 8, are sections, will probably be useful.

This furnace is suitable for a number 30 or 35 black lead pot, with charcoal as fuel. There is no heavy and expensive cast-iron plate on the top, and the cover may be of sheet-iron, although cast-iron is better.

The interior may be cylindrical, but is better, as shown in the figures, narrowed toward the top and bottom. The grate bars rest on an iron ring, Figure 3, supported by an offset in the masonry. The latter is mostly of a very rough kind, as the furnace is sunk in the ground to within a foot of the top, which, besides affording great convenience in working, obviates the necessity of iron bands or stay-rods to support it against the expansion caused by the heat. The greater part of the lining, 4 inches thick, is of good clay, very slightly moistened, and beaten round a hollow wooden core, for which part of a barrel of suitable size answers very well. It is topped with a course of common brick, and above these is a flat iron ring, Figure 4, with lugs, which may be bolted down, or simply let in flush with the top of the furnace.

The grate bars are not built in, but are free to be removed whenever desired. The furnace is constructed, when convenient, at the base of the main smoke stack, with which it is connected by a flue; otherwise a large stove-pipe will answer for a chimney.

When the weight of melted metal does not exceed 100 pounds avoirdupois, the pot is lifted out of the furnace by hand, by means of the basket tongs, the mould being placed on the platform of masonry which is seen in the figure on one side of, and level with the furnace. The melter stands on the top of the furnace to lift, then steps to the ground one foot lower, to pour.

If larger quantities of metal are melted at once, a lever may be employed for lifting the pot. The lever is arranged like that of a blacksmith's bellows, being supported by a rope, chain, or swivel, in such a manner as to admit of a lateral as well as a vertical movement, and at a height of not less than six feet above the furnace, the center of which is directly under the shorter arm.

From the longer arm of the lever a rope depends; from the shorter arm a link of half-inch iron, long enough to reach, when drawn downward, into the cavity of the furnace, and terminating in a hook engaging in an eyebolt which forms the pivot connecting the jaws of the tongs.

The tongs being adjusted on the pot, a ring is slipped over the handles to hold them together; the melter steadies them, while an assistant, pulling on the rope at the other end of the lever, lifts the pot out of the furnace, and swings it near to the mould, when the melter pours the metal.

The hands and arms of the melter are protected by gloves, which, in some works, are elaborately made of canvas and padding. To make a glove, I simply take an ore sack, double it lengthwise, and sew it so as to form a narrow bag, of two thicknesses of canvas, into which the arm may be thrust to the shoulder. The gloves may be wetted to prevent burning, but hot articles must not be grasped with a wet glove, because the steam produced will scald the hand; yet, moisture is a good protection against radiant heat, while grasping the cool handles of the tongs.

In melting with charcoal, the best result is obtained, not by keeping the furnace full, but by letting nearly all the fuel burn away, before refilling. A little practice will enable the melter so to manage that there shall be but little coal in the furnace when the time for pouring arrives, so that it is not in the way when seizing the pot with the basket tongs. The lifting must be performed without delay, otherwise the tongs may become red hot, and bend. The feet and legs of the melter may be protected by woolen armor, or by wetting the boots and trowsers; but melters and assayers must not shrink from a little scorching.

If there is much fuel left in the furnace when the melting is ended, the grate is taken out, and the embers fall into the ash pit, where they are extinguished.

Clinkers, if formed on the sides of the furnace, from the melting of the lining, or from dirty fuel, are punched off, while red hot, by means of an iron bar with a chisel end. If they were to be removed after cooling, the walls would be broken.

If making a bar of clean skimmed metal, it is proper

to throw a little resin, or powdered charcoal into the pot, a few seconds before pouring. It prevents the sputtering which is often caused by the formation of base oxides. As soon as such a bar is cast, the top of it is covered with charcoal powder, to prevent oxidation while solidifying When the slag or matte is poured with the metal the charcoal is not required.

The addition of sand in melting an impure precipitate, especially when, by over-roasting, a good deal of base metal sulphate or oxide has been formed, is very beneficial in saving the pot, by slagging the oxides by which it would otherwise be attacked.

Charcoal is used to effect the deoxidation of sulphates, which would otherwise take place at the expense of the plumbago pot. It is also employed to prevent the formation of a crust, or to reduce the same if formed, on the surface of the matte.

————

SODIUM HYPOSULPHITE.

199. Sodium hyposulphite may be made by passing sulphurous acid (sulphurous anhydride) and air through a solution of sodium sulphide, produced by fusing together five parts of sodium sulphate (a common product of the marshes in the interior) and one part of pulverized charcoal, and extracting with water.

The sodium polysulphide may also be used. It is made by boiling a solution of caustic soda on excess of sulphur, or by fusing together sodium carbonate and sulphur, and dissolving in water. Caustic potassa (concentrated lye) may be used in place of caustic soda, making potassium polysulphide, and then potassium hyposulphite with sulphurous acid and air. The potassium salt answers the same purpose as that of sodium.

Sodium hyposulphite may also be made by heating a solution of sodium sulphite with sulphur. The latter is dissolved. The sulphite is made by passing sulphurous acid through a solution of sodium carbonate, or by exposing the latter, in a moist condition, to the fumes of burning sulphur (sulphurous anhydride) in a reverberatory furnace, or in the flue of a roasting furnace.

For all the purposes mentioned, the sodium carbonate or bicarbonate may be used indifferently.

It has been asserted, by some writers, that the solution of sodium hyposulphite does not dissolve any gold, even in those cases in which the calcium salt does so, referring to that lower gold chloride which sometimes exists in ore which has been roasted with salt (30). This however is a mistake, as has been proved by direct experiment, more than half of the contained gold having been extracted from a sample of such ore, by leaching it with the solution in question.

———

THE CHLORINATION ASSAY.

200. The so-called chlorination assay, is so far incorrect that it gives not only the silver chloride, but all other silver compounds which are soluble in solutions of the alkaline or earthy hyposulphites It may seem that this is a matter of indifference, since all the silver that can thus be extracted can be obtained in the large operation, and this view is correct as to the silver sulphate.

But, in roasting silver ores, if all the requisites of a thorough chloridation are not present, certain compounds of silver are formed, which, though soluble in hypo (or in hot brine), and therefore extracted in the ordinary chlorination assay, are neither chloride nor simple sulphate. These compounds render the extraction of the silver more difficult, and affect the purity of the resulting bullion, whether the leaching process, or amalgamation be employed.

As ammonia dissolves the chloride and sulphate, but not the compounds alluded to, I suggest that comparative assays be made, occasionally, with hypo and with ammonia, as the solvents.

The following account of some investigations, made in 1876, and published at the time in the *Mining and Scientific Press*, will assist in explaining this matter:

1.—Roasted ore containing silver, per ton......$65.00
   A. Leached with ammonia, retained per ton..13.20
   B. Leached with solution of sodium hyposulphite, retained per ton................ 5.20

2.—Roasted ore containing, per ton.......... 60.00
   A. Leached with ammonia (after digestion for several hours in warm ammonia), retained per ton............................. 15.00
   B. Leached with sodium hyposulphite, retained per ton...................... 3.50

3.—Sample of amalgamation tailings containing silver, per ton....................... 9.00
   A. Washed on filter with cold water; filtrate contained a chloride, but gave no precipitate with solution of potassium sulphide.

8

B. Rewashed with hot water; filtrate contained neither chlorine nor heavy metal.

C. Again washed, with cold solution of sodium hyposulphite; filtrate tested with potassium sulphide gave a copious brown precipitate containing silver, lead, arsenic, iron, etc.

D. Excess of potassium sulphide was added to filtrate *C*, and afterwards, to prevent re-solution of arsenic, iron protosulphate in slight excess, then nitric acid to decompose the hyposulphites. A white precipitate was slowly formed, which on boiling became yellow (sulphur).

E.—Precipitate from *D*, removed by filtration, and filtrate treated with an acid solution of silver nitrate, *gave no precipitate.*

The inference is, that the metal salts contained in the washed tailings, and dissolved by solution of sodium hyposulphite, *were not chlorides.*

4.—Sample of amalgamation tailings containing silver, per ton $4.00

A. Washed on filter with hot water; filtrate tested with barium nitrate showed the presence of a sulphate. Washing continued until no more sulphate could be detected.

B. Washing repeated with hot solution of sodium chloride (free from sulphate); a portion of the filtrate tested with potassium sulphide, gave a copious pecipitate; the

remainder tested with barium nitrate also
gave an abundant precipitate.

The inference is that the metal salts extracted by
sodium chloride were sulphates, and as the silver is extracted by hot brine about as perfectly as by sodium
hyposulphite, it seems probable that the metals
exist as a multiple sulphate, otherwise the silver sulphate would have been extracted by hot water alone.
The tailings gave no appreciable silver chloride by
treatment with ammonia. The metals extracted by
hot brine are very slowly reduced by metallic iron, so
that a good result is obtained from ore in this state, in
pans, only by long continued working, and a base bullion is obtained, while the addition of lime to the pulp,
intended to produce finer bullion, entirely prevents the
reduction, causing richer tailings. With silver chloride, even a great excess of lime does not impede
reduction, nor prevent amalgamation, though the latter
is somewhat impeded. By longer roasting of the ore,
the formation of silver chloride was improved, but
the ore required an addition of sulphur, in some form,
to give the best attainable result.

CHLORINATION TAILINGS.

201. However carefully the ore may be treated,
a certain portion of gold remains in the tailings,
varying, other things being equal, according to the
character of the ore. In case this results, in part, from
the presence of particles too large to be entirely dissolved, that portion can be extracted by re-treatment,

either with gas, by chlorine water, or by amalgamation. But a portion, which is sometimes quite considerable, still remains, and resists every mode of chlorination or amalgamation, and can only be extracted by smelting with lead.

In general, it may be said that gold which can be seen by the aid of a lens, after finely grinding, and care- fully washing, a sample of the tailings, can be extracted by chlorination, or amalgamation; but that which can- not thus be rendered visible can neither be chlorinated nor amalgamated, even if the ore be re-roasted, with or without re-grinding. (If re-ground it forms a pasty mass, which cannot be leached, being almost impervi- ous to water.) Of course the ore is supposed to be properly roasted in the first instance.

Concentrated sulphides containing gold, and free from lead, will frequently yield as much as 98 per cent of the fire assay, if moderately rich, but the extraction of 95 per cent is considered a good result, from ore containing $100 worth of gold in a ton. As the mate- rial loses about 24 per cent of its weight in the roasting and leaching, it will readily be perceived that, if there has been no sensible loss of gold in the roasting, the tailings from such ore will assay about $6 per ton, and for every hundred tons of ore treated there will be 76 tons of tailings, containing $456 worth of gold.

Tailings of this character, consisting chiefly of iron peroxide, make an excellent flux for the smelting of galena. All the precious metal they contain is ex- tracted, together with the lead set free from the galena by the action of the iron oxide, and may be considered as clear profit to the smelter, who is obliged to use some kind of iron flux, and can find nothing, unless it be metallic iron, better adapted to his purpose than these tailings. Chlorination tailings can also be utilized in the manufacture of red paint.

———

VALUE OF BARS.

202. Large bullion scales, and troy weights, are not always at hand; yet it is often desirable to know the value of a bar before sending it away from the works, and as good counter scales, with advoirdupois weights, are generally accessible, the following facts may be found useful.

The assay value of one advoirdupois pound of pure silver is \$18.85; that of one advoirdupois ounce is \$1.17 The pound of pure gold is worth \$301.44, and the ounce \$18.84. Hence it is easy to calculate the value of a bar of which the fineness is known, and which has been weighed on common counter scales to the nearest quarter ounce. For example: a silver bar is .969 fine, and weighs 82 lbs 3¼ ounces.

Then: $18.85 \times .969 = \$18.26 \times 82 = \$1,497.32$
and $1.17 \times .969 = 1.13 \times 3\frac{1}{4} = 3.67$

———

Making the value of the bar   \$1,500.99

Each $\frac{1}{1000}$th of silver in a bar is worth 1.885 cents, and each $\frac{1}{1000}$th of gold 30.14 cents per advoirdupois pound; therefore, if in the above example the bar contained also a little gold, say .003, and perfect accuracy is not required, it would be sufficient to add to the value 3 times 30.14 cents for each pound weight, or in all \$74.14. Again, a gold bar is .969 fine and weighs

7 lbs. 13½ oz., or 125½ oz.; then $18.84 \times .969 = 18.26 \times 125\frac{1}{2} = \$2,291.63$. Or, the value of a gold bar may be calculated as though it were silver, and the result multiplied by 16.

The commercial or market value is different from the assay value. The discount on silver varies from 10 to 15 per cent of the assay value, to which is added one-half of one per cent for mintage, or assayer's fee. Gold bars free from base metal are subject to a discount or a premium, equal to one-tenth of one per cent for every ten "points," or thousandths, above or below the quoted par fineness; discount if above, premium if below. The apparent anomaly is due to the fact that, if there is no base metal in the bar, all that is not gold is silver, whence at par fineness, there is just enough of silver in the bar to pay parting charges, etc., while in a finer bar those charges must be met by a discount on the value of the gold. Conversely, a bar which is below par fineness commands a premium for the extra silver it contains. On account of this arrangement, bars which are more than half gold, are stamped with the gold value only, unless they contain much base metal, when both gold and silver value is stamped on them.

Mixed bars in which the gold forms a large part of the value, but less than half the weight, are called "doré" bars, and are stamped with the value of both the gold and the silver.

CHLORINE.

203. Chlorine plays such an important part in connection with human industry, that a short sketch of its

history and properties cannot but be interesting and useful.

Chlorine is an elementary body, as far as is yet known. It is true that this substance has been suspected by chemists to be a compound, but although its decomposition has been more than once announced, it has not, within the writer's knowledge, been verified.

Chlorine was discovered by Scheele in 1774, and, in accordance with the crude notions then entertained, was called "*dephlogisticated marine air.*" Sir Humphrey Davy investigated its properties at a later date, and gave it the name which it now bears, derived from a Greek word signifying pale green, or yellowish green. The name of the mineral "chlorite" has the same derivation, but this substance contains no chlorine, and must not be confounded with the chlorous acid salts, which are also called chlorites.

Under ordinary circumstances, chlorine is a pungent and suffocating gas of a yellowish green color, and is 2.44 times as heavy as an equal volume of air. When compressed to one-fifth of its normal volume it becomes liquid, and is then 1.33 times as heavy as an equal volume of water.

Chlorine has powerful affinities, which cause it to combine, under suitable conditions, with almost every other known element. When naturally combined with silver, it forms the well known mineral *horn silver*. It is also found in nature, combined with lead, copper, and other metals, especially with sodium in the compound known as common salt. When it is considered that the entire ocean, covering three-fourths of the surface of the globe, with a depth, in places, of several miles, is heavily charged with salt, and that over six-tenths of the weight of salt consists of chorine, it will be perceived that a more definite statement could hardly convey a more definite idea of the enormous quantity of chlorine existing in the world. If all the chlorine in the ocean could be set free at once, it would destroy every vestige of life on the globe.

In combination with hydrogen, chlorine forms hydrochloric acid gas, and in this form is evolved in enormous quantities in the manufacture of soda-ash from salt, by the agency of sulphuric acid. The gas, dissolved in water, forms ordinary hydrochloric or muriatic acid. This acid, in contact with manganese binoxide, evolves chlorine, and it is in this way that chlorine is produced, in countries where the acid is a waste product of the soda works, for the manufacture of the so-called chloride of lime, or bleaching powder. In this country the use of chlorine is limited, and we generally make the hydrochloric acid, and develop chlorine from it at the same time, by means of a mixture of salt, manganese binoxide, and dilute sulphuric acid, as explained in the directions for the chlorination of gold ores. The best remedy for the effects of an accidental inhalation of chlorine is, to breathe the vapor of water or alcohol, or to drink a glass of spirits.

PLATTNER'S PROCESS.

204. The process of gold extraction known as Plattner's chlorination process, which is that described

in the text, originated in Europe, and was first introduced here by G. F. Deetken, M. E., at Grass Valley,
in Nevada County.

In 1866, Mr. Deetken having removed from Grass
Valley, the operators who succeeded him were suddenly
confronted by a difficulty in the treatment of the concentrations from the Eureka mine, which all at once
developed an amazing capacity for absorbing chlorine,
so that the application of Plattner's process to them
was no longer profitable. The writer, happening to
visit Grass Valley in that year, was applied to for a
remedy for the difficulty, which, not being aware of the
cause, nor having facilities for investigation, he was not
able to supply; but the matter being made public
through the medium of the *Mining and Scientific Press*,
it arrested the attention of Mr. Deetken, who at once
repaired to the spot, and, with his well known skill, and
professional sagacity, detected the cause of the difficulty and applied the remedy.

The trouble was occasioned by the presence of a
magnesian compound in the concentrated sulphurets.
By the addition of a small quantity of salt to the roasting mass, this substance was converted into magnesium
chloride in the furnace, much more cheaply than it
could be done by means of chlorine in the vats. The
gold was then chloridized in the usual manner. The
use of salt in the furnace for this purpose was, so far as
is known to the writer, original with Mr. Deetken, being
quite distinct from its use for the purpose of chloridizing silver, when present in the ore. To this gentle-
man we also owe the demonstration of the possibility of
conducting the process in wooden vessels.

THE KISS PROCESS.

205. The use of alkaline or earthy hyposulphites for the extraction of silver from ores, also originated in Europe, having been suggested by Dr. Percy, in 1848, and first applied in practice by Von Patera, in 1858, at Joachimsthal. Von Patera used the sodium hyposulphite for the leaching, and the sodium polysulphide for the precipitation. This method was modified by Kiss, who substituted the calcium salts for those of sodium.

The Kiss process was first successfully introduced on this coast by Kustel and Hofman, in Mexico, after abortive attempts had been made by others in Lower California. It met with such success as to supercede amalgamation in the treatment of silver ores, throughout the State of Sonora, and the Territory of Lower California, and was even known among ill informed people as the " Mexican process." It has been gradually growing in public favor, the principal obstacle to its extensive adoption in the silver regions of Nevada being the circumstance that the ores of that State usually carry more or less gold, which, while frequently not justifying the application of the Plattner process, is but partially, and almost by chance, extracted in the Kiss silver process.

One great objection to the use of the Plattner process in these cases is, the cost of acid, the transportation of which is troublesome and expensive. It is therefore desirable that some means of developing pure chlorine from salt cheaply, without the use of acid, should be devised.

THE BRÜCKNER FURNACE.

206. This furnace consists of a horizontal, brick-lined, hollow cylinder, of boiler iron, with central openings at the ends, through which the flames from a fixed fireplace pass to a flue connecting with a dust chamber, while the cylinder rotates slowly on rollers. The ore is introduced and discharged, periodically, by means of trap doors in the side of the cylinder. The latter is divided lengthwise by a reticulated, spiral partition, or diaphragm, composed of cast-iron plates supported by hollow cast-iron bars, which, passing through the walls of the cylinder, admit of a circulation of air through them, with the design of protecting them from too great heat.

The function of the diaphragm is twofold; firstly, to lift the ore, and shower it through the heated air within the furnace as the latter revolves; secondly, to move the ore from end to end of the cylinder, in order that all portions of the ore may be equally exposed to heat.

The diaphragm is liable to rapid destruction by the action of sulphur, arsenic, and antimony in the ore. It is stated, on excellent authority, that the loss of the diaphragm does not impair the efficiency of the furnace. In some cases the furnace has been found to work better without the diaphragm, forming fewer lumps in the roasting of leady ores.

In this, as in all rotating cylinder furnaces, the light and fine portion of the ore is carried away by the force of the draft. It is therefore necessary that extensive dust chambers be provided, through which the draft from the furnace must pass, on its way to the chimney, in order that the principal part of the dust may be deposited.

The writer superintended the erection and working of one of these furnaces in Inyo County, and obtained satisfactory results from it. About 15 per cent of the ore passed to the dust chamber, and was found to be but imperfectly roasted. This dust was collected, slightly moistened with a solution of salt and iron sulphate in water, and then returned to the furnace, and successfully roasted. About the same percentage of it again found its way to the dust chamber.

The cylinder is twelve feet long by six feet in diameter, and a charge for it is from $1\frac{1}{2}$ to $2\frac{1}{2}$ tons of ore, with the necessary salt. The time required for the roasting of a charge of ore in this furnace is from four to twelve hours, including the charging and discharging.

As the weight of the charge varies with the quality of the ore, and the degree of fineness to which it is crushed, and the time required for the roasting varies with the same circumstances, the capacity of the furnace may be said to range from four to twelve tons in 24 hours. One man on a shift can do all the work, or can attend to the firing of four furnaces.

This furnace has been extensively used in treating the silver ores of Colorado and Mexico, and one, without the diaphragm, is now, or was recently employed in the roasting of auriferous sulphurets in Nevada County, California.

THE BRUNTON FURNACE.

207. This furnace is essentially similar to the Brückner, but has no diaphragm. In form it differs slightly, being ovoid, or egg shaped, the larger end toward the fire. This form must conduce to the stability of the brick lining, and is said to roast the ore more evenly than the cylinder. The rotation of the furnace is not produced, as in the Brückner, by means of a pinion engaging in a toothed rack surrounding the shell, but simply by the traction of the rollers, two of which are connected by a strong shaft, to which motion is imparted by means of a screw on a transverse shaft. This, with the absence of the diaphragm, a low royalty for the patent right, and some other peculiarities, render the furnace cheaper than the Brückner. A Brunton furnace twelve feet long by six feet eight inches in its greatest diameter, recently constructed for the chlorination works at the Plumas National Mine, in Plumas County, California, weighed, including gearing and iron bed-plates for the rollers, in place of the wooden ones formerly used, nearly 10 tons without the lining bricks.

THE PACIFIC CHLORIDIZING FURNACE.

208. This is a simple brick lined rotating cylinder, not differing essentially from those described. It is however made of large capacity, as much as seven tons of ore being treated at once. A good feature in the make-up of this furnace is that the charging hoppers,

of which there are two, are supplied by the foundry already mounted on iron supports.

THE WHITE FURNACE.

209.   The White furnace consists of a hollow rotatory brick-lined cylinder of cast-iron, with open ends.   The brick lining is so arranged, in the short segments of which the cylinder is composed, as to form "grooves, cavities or projections," by which the ore is lifted and showered through the flames, which pass through the furnace from end to end.   The cylinder is slightly inclined, either downward from the fireplace, or the reverse, and the pulverized ore is poured automatically and continuously, together with the salt, into the higher end.   The rotation, aided by the lifting and dropping of the ore, causes the latter to traverse the entire length of the cylinder, and to fall continuously from the lower end.   By an ingenious arrangement, the inclination of the cylinder is rendered adjustable, so that the ore may occupy a longer or a shorter time in passing through it.

When the furnace is so arranged that the cylinder is inclined upward from the fireplace, and the ore consequently enters at the cooler end, it results that the lighter dust does not pass through the furnace, but is carried by the force of the draft in the opposite direction, toward the flue, and into a series of chambers in which the greater part of it settles, while the smoke, etc., passes on to the chimney.   To effect the roasting of this portion of the ore, an auxiliary fire is necessary, through the flames of which the dust may pass on its way to the chambers.   The coarser portion of the ore passes along the cylinder, under continually increasing heat, and, on its exit at the lower end, falls into a pit, situated between the end of the cylinder and the main fireplace, where it is exposed to the heat of the flames passing over it.   At stated periods, the accumulated ore is removed from the pit and spread on the cooling floor.

This arrangement of the cylinder—the ore entering at the cooler end—fulfils, better than the reverse plan, the requisite conditions of a proper roasting.   The writer is not aware of an instance in which the inclination was downward from the fireplace, and the ore entered the furnace at the hotter end, in which satisfactory results were obtained.

THE HOWELL–WHITE FURNACE.

210.   The furnace known by the above appellation, is a modification of the White.   The cylinder is lined with bricks in only one-third of its length from the lower end, adjoining the fireplace, the diameter of this portion of the iron shell being 10 inches greater than that of the unlined portion

The ore enters the cylinder at the upper and cooler end, and is lifted and showered by means of projections which in the unlined portion are formed of iron, in the lined portion of projecting bricks.

The inclination of the cylinder is fixed, the time occupied by the ore in passing through it being regulated by the number of rotations per minute, which is adjustable at will.

The length of the cylinder is 23 feet 2 inches. The internal diameter of the smaller part of the iron shell is from 22 to 50 inches, that of the enlarged portion being as stated 10 inches more, in order to admit of lining.

The working capacity of the furnace ranges from eight to forty-five tons of ore in 24 hours, according to the character of the ore and the diameter of the cylinder.

―――――

THE THOMPSON–WHITE FURNACE.

Another modification of the White furnace has been made by J. M. Thompson.

This furnace differs from the Howell-White in several particulars. The cylinder is of one diameter throughout its length, and is lined throughout with tiles, thus lessening the loss of heat by radiation, which must be considerable in the unlined portion of the Howell-White furnace. The tiles are specially molded for the purpose, and are so formed that the projections and recesses, designed to lift the ore, are of a rounded form as to their transverse section. In some cases the recesses are obstructed at intervals by transverse ridges. The intention of this is to prevent the ore sliding along in the inclined cylinder, in consequence of its low angle of stability at an early stage of the roasting. In some ores this angle becomes almost *nil*, so that the powder resembles a fluid in its action.

The further retention and equalization of heat in the cylinder, is promoted by means of a layer of non-conducting material, such as paris plaster, or asbestos, between the lining, of bricks or tiles, and the iron shell.

The latter, and the supporting rollers with their bear-ings, are thus, in a measure, protected from the injuri-ous effects of unequal or excessive heat, by being kept comparatively cool.

The furnace is arranged in either of two ways. In the one, the whole of the ore enters the cylinder at the higher and cooler end. In this case the lighter dusty portion which is carried back by the draft, is roasted by the flames from an auxiliary fire between the end of the cylinder and the dust chambers, the latter being constructed of masonry. In the other, the dust is sep-arated from the coarser portion of the ore before enter-ing the cylinder. The coarse portion then enters at the upper end, while the dust enters at the lower end near the fire. The main fire is thus made to serve for the roasting of both portions of the ore, and the auxil-iary fire is dispensed with. It is also claimed by the inventor, with considerable show of reason, that the heat and the chlorine, developed by the roasting of the dust, mingled as it is with salt, instead of being lost, as when an auxiliary fire is relied on, aids materially in chloridizing the coarser portion of the ore.

A portion of the dust, impelled by the draft, traverses the entire length of the cylinder, and entering the flue is arrested in its flight by the ordinary means of dust chambers, which, in this case, are constructed of iron instead of masonry; hence they are light, portable, and cheap.

The length of time occupied by the passage of the coarser portion of the ore though the cylinder, is regu-lated mainly by the inclination of the latter, which, as in the White furnace, is adjustable, though, as the writer is informed, in a different manner. Mr. Thompson prefers to regulate the period of transit by varying the angle of inclination, rather than the speed of rotation, of the cylinder, arguing that, as it is stated by eminent metallurgists that the loss of silver by volatilization is, other things being equal, proportioned to the length of time during which the ore is exposed to the heat, it follows that the more quickly the ore is roasted the less loss it will sustain. But, he continues, the greater or less rapidity of the roasting depends on the greater or less exposure of the particles of ore to, first oxidizing, then chloridizing influences, or to both simultaneously; hence the more continuously the ore is showered through the heated air and gases pervading the interior of the furnace, the more quickly will the roasting be effected, and consequently less loss will be caused by the volatilization of silver.

This theory seems plausible. Let us examine it. Let us suppose two furnaces in operation, receiving equal quantities of the same kind of ore. They are alike in every respect, rotating with equal speed, fired in the same manner, and with the same amount of draft. They will yield like results.

We will now suppose that both of the cylinders are rotating at the highest feasible rate of speed, thus affording the greatest exposure of the ore to the roasting influences. It is desired to modify their action so that the time occupied by the passage of the ore through them may be doubled.

In one, which we will call the Howell, this result is attained by reducing the speed of rotation. (Owing to the effect of tangential force the reduction of speed will be somewhat less than half.) In the other, or Thompson, by lessening the angle of inclination. The supply of ore to each is supposed to remain unchanged for the present. An analysis of the effects of these alterations gives the following results.

The quantity of ore passed through each furnace in a given time remains unaltered.

The quantity of fuel and air consumed by each is unchanged.

The quantity of ore exposed to the heat at any given moment in each is doubled; therefore, the time during which the ore is exposed to heat in each is doubled.

The length of time during which the ore is dropping through the air is, in the Howell, unchanged; in the Thompson, doubled.

The power consumed in lifting the ore is—in the Howell unchanged; in the Thompson, doubled.

In the Howell the same quantity of work is performed, on the same quantity of ore, with the same expenditure of power, as before the change. In the Thompson a double quantity of work is performed on the same quantity of ore, by a double expenditure of power. If the ore is as well roasted in the one furnace as in the other, the advantage is, so far, with the Howell. Is it probable that the same result can be produced by the Howell as by the Thompson, with half the work, and therefore half the power? As the result sought is a chemical one, the answer to this question depends on the extent to which the attainment of that result is accelerated by the movement of the ore.

The time occupied by the oxidation of a single particle of sulphuret, exposed to heated air, other things being equal, is in some ratio to its mass. It matters not whether an adjacent particle consists of sulphuret or quartz; the two particles of sulphuret will be oxidized simultaneously, in the same length of time as would one particle, *provided the requisite quantity of heated oxygen be supplied.* As regards the heat, the two particles will aid each other by their combustion, rendering necessary a smaller proportionate quantity of fuel than if one particle were sulphuret, and the other quartz.

In a roasting cylinder there is, at any given moment, a certain quantity of air, capable of oxidizing a certain number of particles of sulphuret. To expose a greater number of particles would be useless. To expend power for that purpose would be to waste the power.

If we suppose the quantity of oxygen passing through each furnace to be in excess of the quantity which can be utilized, the feed may be increased in both. When the quantity of air passing through the furnaces is in equilibrium with the ore, so that the oxygen is utilized to the utmost practicable extent, one of two things must be true. Either the Thompson is wasting power in lifting the ore more often than the Howell, or, the ore is remaining in it a longer time than is necessary. If the former, the rate of rotation should be reduced; if the latter, the inclination of the cylinder should be increased. The latter is the most probable, because, although it is not denied that oxidation proceeds to a certain extent while the ore lies quiescent in the cylinder, it is proved, by all experience, to proceed faster when the ore is showered through the heated air, provided, as before said, that a sufficient quantity of free oxygen is present, which, on our hypothesis, is the case in this instance.

If the speed of the Thompson is now increased, the same quantity of ore will pass through it, but the time occupied in the passage will be lessened, and the load in the furnace will be reduced. Less power will now be consumed in causing the cylinder to rotate, and the loss of silver by volatilization will be diminished.

Whether the power required would be reduced to an equality with that consumed by the Howell, which would mean that the load in the furnace should be reduced to half that in the Howell, or, in other words the time of passage should be half, or whether the result would be a mean in which the diminution of volatilization would compensate, or more than compensate for the greater consumption of power, is a question which can only be answered by experiment.

A similar view as to the chloridation, which, in this class of furnace, is more or less effected by the falling of the ore particles through the heated gases evolved by the roasting ore and the salt, would lead to similar conclusions. But the theory of the chloridation is opposed to any greater movement of the ore in this stage than is necessary to ensure equal and thorough heating. Hence it would perhaps be advantageous to have a smaller number of lifters in the circumference of the cylinder near its lower or chloridizing end, than in the central and upper zones. But it is known that, in this class of furnace, a considerable percentage of the chloridation takes place after the ore has left the cylinder, and while it is accumulating in the pit into which the cylinder empties itself. Exactly how far this principle may be carried in practice has not been proved.

The objection that rapid rotation of the cylinder causes the separation of a larger quantity of dust from the roasting ore, is too puerile to be worth refuting. The only result is that the auxiliary fire has more to do, and the cylinder less, a clear saving of motive power, or an increase in the capacity of the furnace. One point in this connection is worthy of notice; the dusting almost ceases as soon as the ore begins to "sponge." This is in favor of feeding the dust into the lower end of the cylinder.

Another valuable feature in the Thompson furnace, and which has been heretofore surprisingly overlooked by inventors, is that the heat given off by the roasted ore, instead of being wasted, is utilized in heating the air with which the furnace is supplied; at the same time the ore is cooled sufficiently to allow of its being moistened without loss or inconvenience from the formation of a great volume of steam.

Thompson also adapts the length of the cylinder in such proportion to its diameter, and to the extent of grate surface in the fireplace, as that the heat at the higher end shall be, as nearly as practicable, that which is proper for the initial stage of the roasting. It is a self-evident proposition that a large quantity of burning fuel will heat a cylinder, not only of greater diameter, but of greater length, than a smaller quantity. Hence, if the length of the cylinder is not increased in suitable proportion to the increased diameter, one of two things must happen: either the larger cylinder, by being too short, must waste available heat at its upper end, or, the smaller cylinder, being too long, is not sufficiently heated throughout. As the proper initial heat varies for different ores, the most advantageous length of cylinder for a given diameter will also vary. The usual dimensions are 21, 24, 27, and 30 feet in length, to 32, 40, 52, and 60 inches in diameter, respectively.

As to the question of the best manner of regulating the operation of a roasting furnace of this class, the writer's opinion is, that the cylinder should be adjustable both as to speed of rotation and as to inclination, in order that it may be adapted to the various require- ments of different ores. In the White, and in Thomp son's modification, this is the case.

The roasting of ores is a problem in which mechani- cal and chemical factors are inextricably interwoven. While it is well that theory should suggest and guide experiment, yet, if the facts thus developed oppose the theory, the latter, not the former, must give way.

THE O'HARA FURNACE.

212. This is a reverberatory furnace of great length, the hearth being about one hundred feet long by five to eight feet wide. It is traversed lengthwise by plows and scrapers attached to endless chains, which, passing over rollers at the ends of the furnace, carry the plows back to the place of entrance, where the ore is fed in continuously by a mechanical arrangement. By the ac- tion of the plows the ore is not only turned over, but is also progressed gradually towards the hottest part of the furnace, thence to a cooling hearth, and is finally

discharged into a suitable receptacle. In a furnace of this size several fires are used on each side, at different points in the length. A great advantage of this furnace is, that very little of the ore is carried to the flues by the draft. It can be operated by one man on a shift, and has a capacity of as much as 40 tons per 24 hours. By a recent improvement the furnace is built in two stories.

———

STETEFELDT FURNACE.

213. If any metallurgist had been told, prior to the introduction of the Stetefeldt furnace, that a chloridizing roasting of silver ore could be effected in the short space of a few seconds of time, he would probably have been incredulous. In the old reverberatory, the operation requires several hours; in the cylinder furnaces of continuous action it occupies from 10 to 30 minutes, but in the furnace now under consideration, the ore, if not of extremely refractory character, is almost completely roasted while falling from a height of little more than 20 feet.

The furnace consists of a stack or shaft of masonry, provided with fireplaces opening into it near the base, with a flue near its upper end, and with an apparatus for sifting the ore and salt into it at the top. The ore thus showered from the top of the stack, encounters in its descent, under the most favorable conditions for chemical action, the ascending flames and heated air from the fireplaces, as well as the fumes of its own combustion, and from the voltalization and decomposition of the salt. When it reaches the bottom, where a hopper is provided for its reception, not only is the oxidation completed, but the chloridation of the silver is so far advanced as to require for its completion only the further exposure to heat which the ore receives while accumulating in the hopper. The latter is emptied from time to time, by means of a slide, into iron cars and the ore removed to the cooling floor.

As in all other processes in which the pulverized ore is showered through the flames, the lighter portion, amounting sometimes to 30 per cent of the whole, is carried back by the current of air and gas. In order to prevent an accumulation of dust in the flue, the latter joins the stack at an acute angle, and is conducted almost vertically downward to the base, where the flames from an auxiliary fireplace enter it, and effect the roasting of the dust, which then, passing into a horizontal flue, settles mainly in a series of hoppers, while the hot air, smoke and gases pass through dust chambers to the chimney. The construction of the furnace involves many ingenious contrivances for regulating the feed and heat, for the removal of the roasted ore, the inspection of the interior, and the admission of air at suitable points, and in proper quantity.

In a large furnace, capable of roasting from 30 to 70 tons of ore in 24 hours, the roasting tower is 25 feet high and five feet square at the base. One man on a shift can attend to the firing.

The following extract from the Circular of the Stetefeldt Furnace Co. will assist, with the drawings, to an understanding of the arrangement:

"Of the accompanying drawings Figure 1 represents a vertical section of the Stetefeldt Furnace, showing its latest and most improvd mode of construction, and Figure 2 is a sketch of the Stetefeldt Feeder.

DESCRIPTION OF THE FURNACE.

A is the shaft into which the pulverized ore is showered by the feeding machine, placed on the top of the cast iron frame B. The shaft is heated by two fireplaces (C). The ashpits of these are closed by iron doors, having an opening (E), provided with a slide, so that more or less air can be admitted below the grate, and, consequently, more or less heat generated. In order to obtain a perfect combustion of the gases, leaving the firebox through the slit (T), an airslit (U), connected with the airchannel (F), is arranged above the arch of the firebox. This slit also supplies the air necessary for the oxidation of the sulphur and the base metals. Another advantage of this construction is that the arches above the firebox and firebridge are cooled and prevented from burning out. The roasted ore acmulates in the hopper (K), and is discharged into an iron car by pulling the damper (L), which rests on brackets with friction rollers (M). N is an observation door, and also serves for cleaning the firebridges. O are doors to admit tools in case the roasted ore is sticky and adheres to the walls. The gases and fine ore dust, which forms a considerable portion of the charge, leave the shaft through the flue (G). The doors (R) are provided to clean this flue, which is necessary, with some ores, about once a month. D is an auxiliary fireplace, constructed in the same manner as the fireplaces on the shaft, which is provided to roast the ore dust, escaping through the flue (G), in passing through the chamber (H). P are doors for observation and cleaning. The larger portion of the roasted dust settles in the chamber (V). provided with discharge hoppers (I), from which the charge is drawn into iron cars by moving the dampers (S). The rest of the dust is collected in a system of dust chambers (Q), connected with a chimney which should rise from 40 to 50 feet above the top of the shaft. At the end of the dust chambers is a damper by which the draft of the furnace can be regulated. The dry kiln can also be used as a dust chamber, and the waste heat of the furnace utilized for drying the ore before crushing it. The firing of the furnace is done on one side, and all discharges are located on the opposite side."

DESCRIPTION OF THE FEEDING MACHINE.

" The Feeding Machine is shown in Fig. 2. The castiron frame (A), which is placed on top of the shaft, is provided with a damper (B), which is drawn out when the furnace is in operation, but inserted when the feeding machine stops for any length of time, or if screens have to be. replaced. C is a castiron grate, to the top of which is fastened the punched screen (D). The latter is made of Russian sheetiron, or of cast-steel plate, with holes of one-eighth to one-tenth of an inch in diameter. Above the punched screen is placed a frame (E), to the bottom of which is fastened a coarse wire screen (F), generally No. 3, made of extra heavy iron wire. The frame (E) rests upon friction rollers (G). The brackets (H) which hold the friction rollers can be raised or lowered by set screws, so that the wire screen (F) can be brought more or less close to the punched screen (D). The brackets (K) carry an eccentric shaft (L), connected with the shaft (M), from which the frame (E) receives an oscillating motion. To the brackets (N) are fastened transverse stationary blades (O), which come nearly in contact with the wire screen (F), and can be raised or lowered by the nuts (P). These blades keep the pulp in place when the frame (E) is in motion, and also act as distributors of the pulp over the whole surface of the screen. The hopper (I) receives the ore from an elevator which draws its supply from a hopper into which the pulverized ore is discharged from the crushing machinery. The ore is generally pulverized through a No 40 screen. By means of a set of cone pulleys the speed of the frame (E) can be changed from twenty to sixty strokes per minute, whereby the amount of ore fed into the furnace is regulated. This can also be done to some extent., by changing the distances between the punched screen (D), the wire screen (F), and the blades (O).

The largest sized furnace, as represented in the drawing—the scale of which is 1 in. = 12 ft., capable of roasting from 50 to 70 tons of ordinary ores, and from 30 to 35 tons of very base sulphuret ores in 24 hours, requires the following amount of materials, from which the cost of construction can be easily calculated by any architect or millwright, viz:

1,500 fire-bricks, for fire boxes and arches exposed to flame.

200,000 common bricks, of good quality, for furnace, large system of dust chambers, chimney, and cooling floor.

2,500 lbs. in bolts and nuts for anchoring furnace and dust chambers.

4,500 lbs. in wrought iron braces, flat iron for car-guides, tools, etc.

16,000 lbs. in castings.

All the castings are very plain and simple, the water-jacket on top of furnace, and the water damper having been discarded. Considerable work is only required on the feeding-machine, feeding-machine damper, and discharge damper, and some on the fire-doors, which will be covered by an additional charge of about $700 added to the ordinary price of castings.

The cost of three iron discharge cars is $125. For a furnace of 15 to 20 tons capacity, without hopper discharge, and a less extensive system of dust chambers, the amount of materials required may be estimated at two-thirds the figures given above."

REMARKS ON FURNACES.

214. The quantity of fuel consumed in roasting varies with the quality of the fuel, the character of the ore, and the kind of furnace used.

In a single hearth reverberatory furnace, each ton of ordinary silver ore requires from one-third to one-half a cord of dry pine wood. In a long furnace the proportion is materially reduced.

Concentrated auriferous sulphides, when roasted for chlorination, in a reverberatory furnace with three hearths, require one cord of wood, or one-half ton of Seattle coal, per ton of ore. In this roasting, one man on a shift is sufficient for one furnace, roasting $1\frac{1}{2}$ tons of ore in each 24 hours. The writer experimented with different numbers of men employed at one time, but found no advantage in more than one for all three hearths, thus proving that stirring the ore beyond a certain extent is useless, unless the supply of air be increased. In the mechanical furnaces the proportionate consumption of fuel is much dimished, especially in those which receive the ore continuously, in which one-tenth of a cord of wood to the ton of silver ore is a common proportion. The mechanical furnaces described are all in more or less extensive use, giving satisfactory results, with great economy of fuel and labor as compared with the old reverberatory furnace.

Those which receive the ore continuously are the most economical in operation, but the most costly in construction. They are especially adapted to the chloridizing roasting of large quantities of silver ore of nearly uniform character.

For the roasting of gold-bearing sulphides, or exceptionally rich silver ores, for custom works in which the character of different small lots of ore is liable to extreme variation, or in small establishments, the furnaces which are charged periodically are usually preferred, because the roasting in them is more readily controlled, or because of their more moderate cost.

The furnaces which are charged continuously, as the

Stetefeldt, and the White with its modifications, have not yet been tried for the dead roasting of gold-bearing sulphurets for the Plattner process, but there is no reason to doubt that, with proper care, they might be made to do the work. Possibly a modification, such as that used by Mr. Crosby (50) in connection with his reverberatory furnace, might be necessary on account of the large quantity of sulphur which the material in question contains.

One disadvantage of the continuous cylinder furnaces in which the ore progresses in an opposite direction to the draft is, that the heavier particles, which require the longest exposure, pass through more quickly than the lighter, because the force of the draft has less influence in retarding their horizontal progression. The action of the furnace must be adjusted to the requirements of these heavy particles. Between these and the dust which is completely controlled by the draft, are particles of all grades, some of which are barely massive enough to make headway against the current of air.

As the time required for the roasting of the larger particles is greater than that required for the smaller, it is clear that a large proportion of the ore which passes through the cylinder is detained within it longer than is necessary, causing useless consumption of power and, in all probability, loss of silver. This defect is obviated when the ore moves through the furnace in the same direction as the draft. The progression of the ore is then aided by the force of the draft, but in a less degree as the particles are heavier. The objection to this method is, that the ore in its progress encounters a gradually diminishing heat, and a decreased proportion of oxygen, while the reverse should be the case.

Mr. White, the original inventor of the application of this kind of furnace to the treatment of silver ores, has been much blamed for his persistence in this manner of working it, but it must be admitted that, if the objection pointed out could be overcome, the method would possess certain advantages.

In the Stetefeldt furnace, in like manner, the descent of the heavier particles is less impeded by the upward current of air and gas, and is therefore more rapid than that of the lighter, but in this case no waste of power results, because the ore is lifted, once for all, to the top of the shaft by an elevator.

In the O'Hara, all portions of the ore, except a small quantity of very light dust, are moved toward the hotter end of the furnace with practical uniformity. This furnace combines the advantages of the reverberatory, worked by hand, with those of automatic action. It probably costs more for repairs than the others.

A roasting furnace, of whatever description, should be kept in operation as constantly as possible, not only in order to economize fuel, but also to avoid the injurious effect of alternate heating and cooling. In the selection, therefore, regard should be paid to the quantity of ore to be treated daily, as well as to its character, and the financial resources at command.

INDEX.

Plate I.

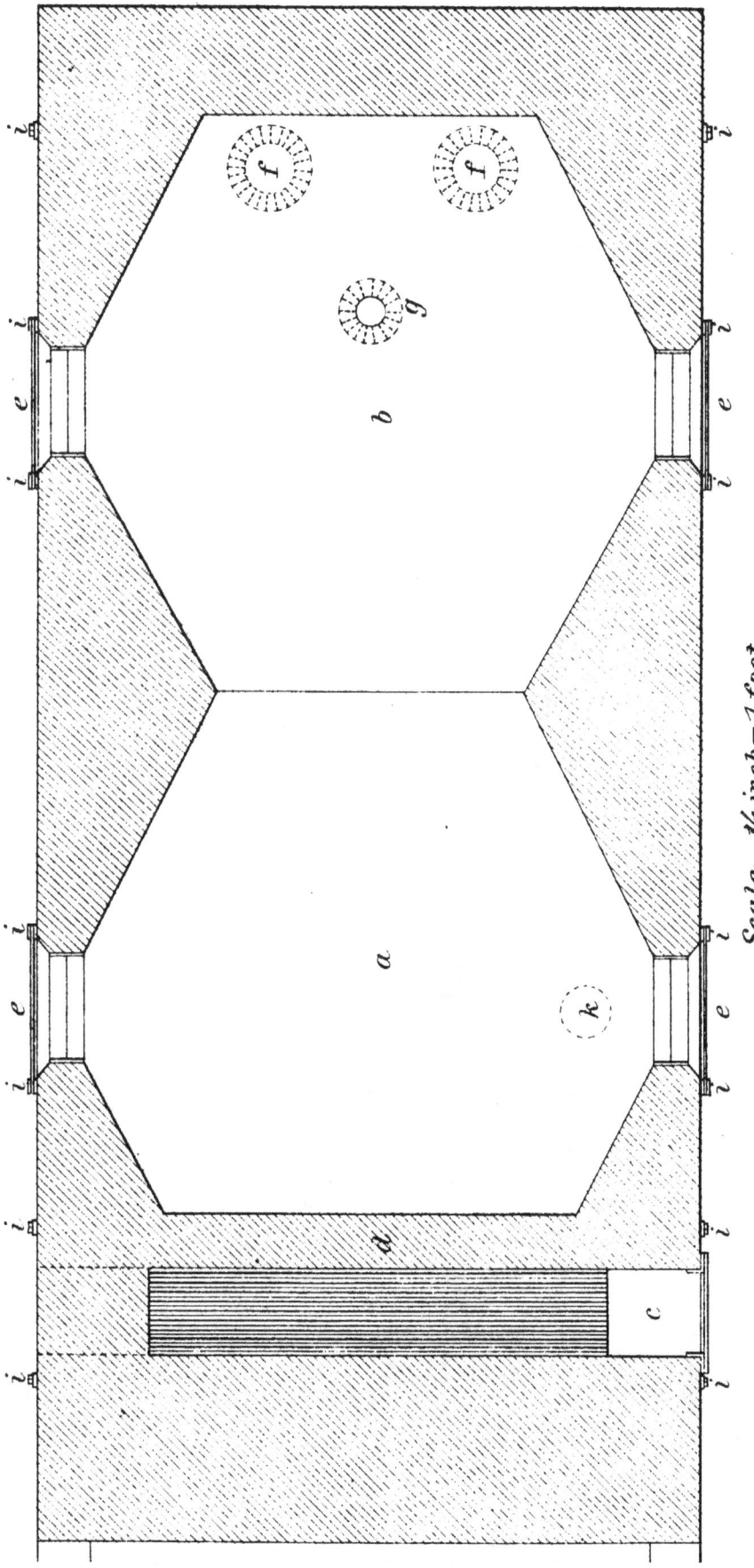

Scale; ½ inch = 1 foot.

a. First hearth. b. Second hearth. c. Fire place. d. Fire wall. e. Doors. f. Flues. g. Charge hole. i. Buckstay. k. Discharge hole.

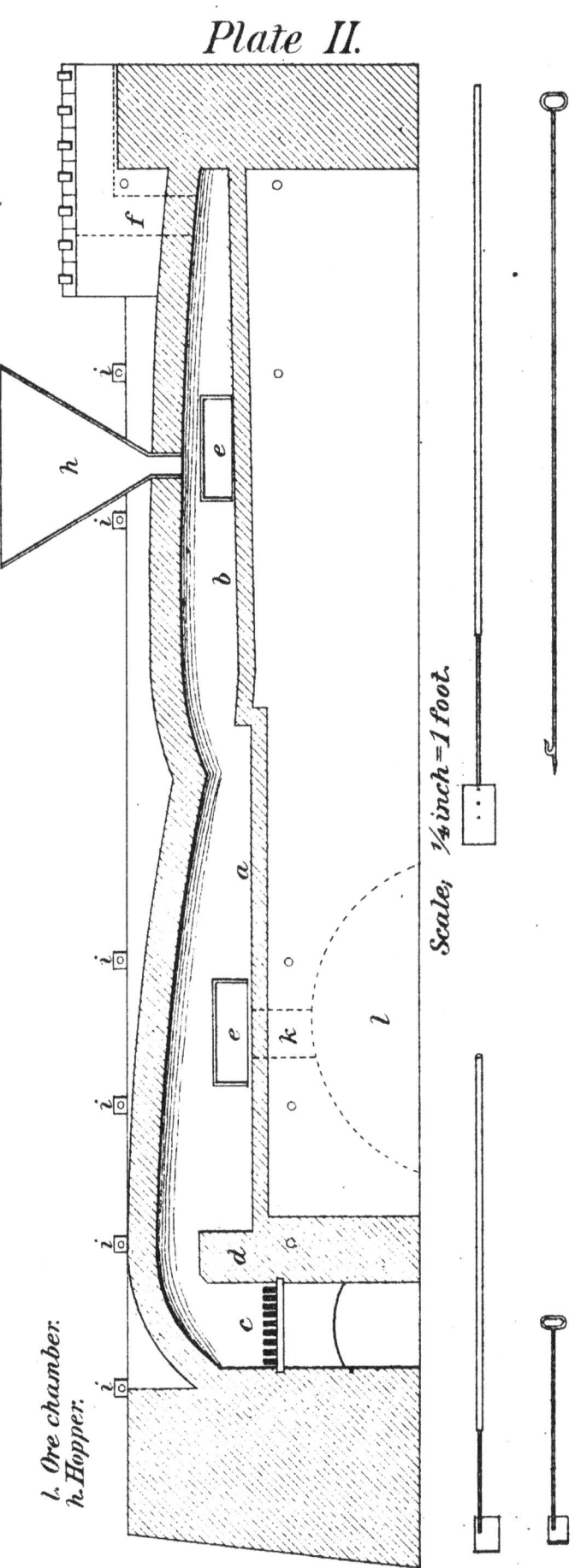

Plate II.

Scale; ½ inch=1 foot.

l. Ore chamber.
h. Hopper.

Plate III.

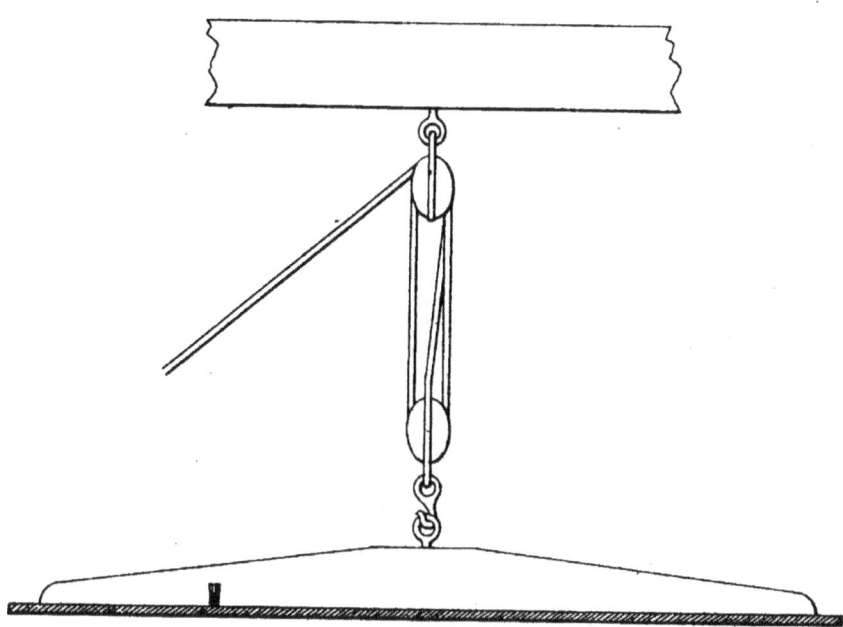

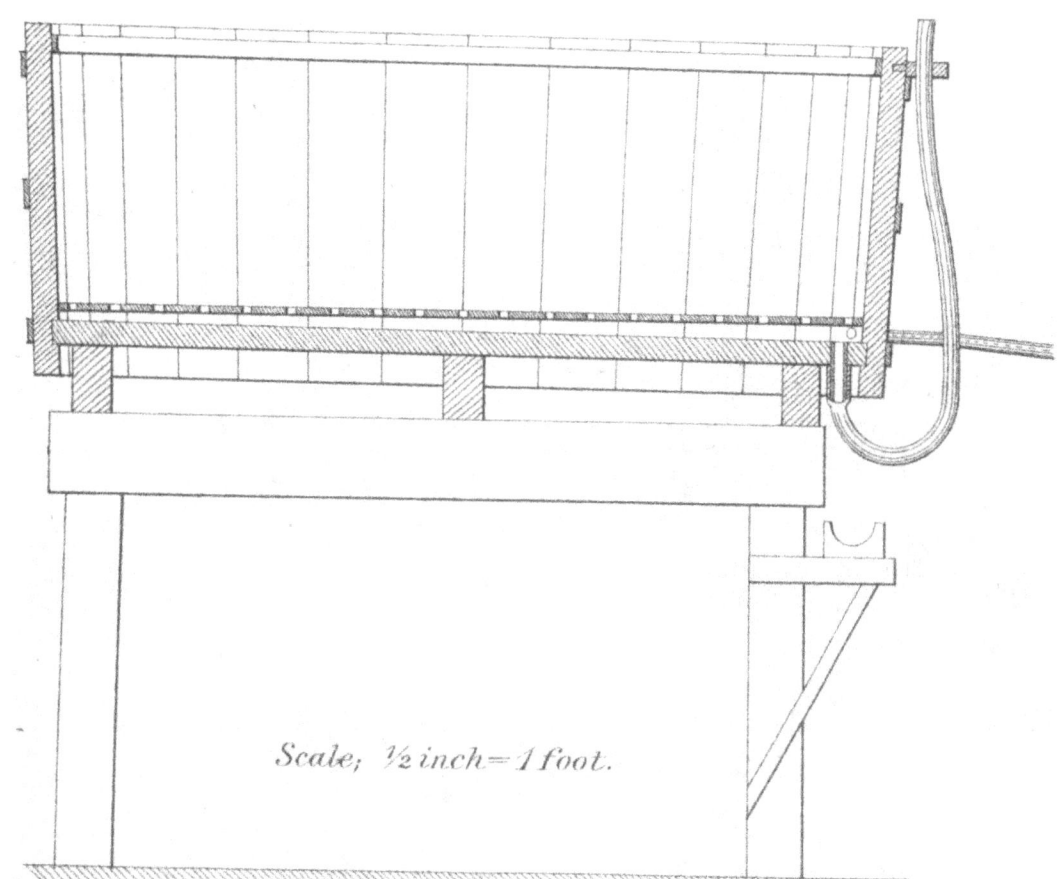

Scale; ½ inch = 1 foot.

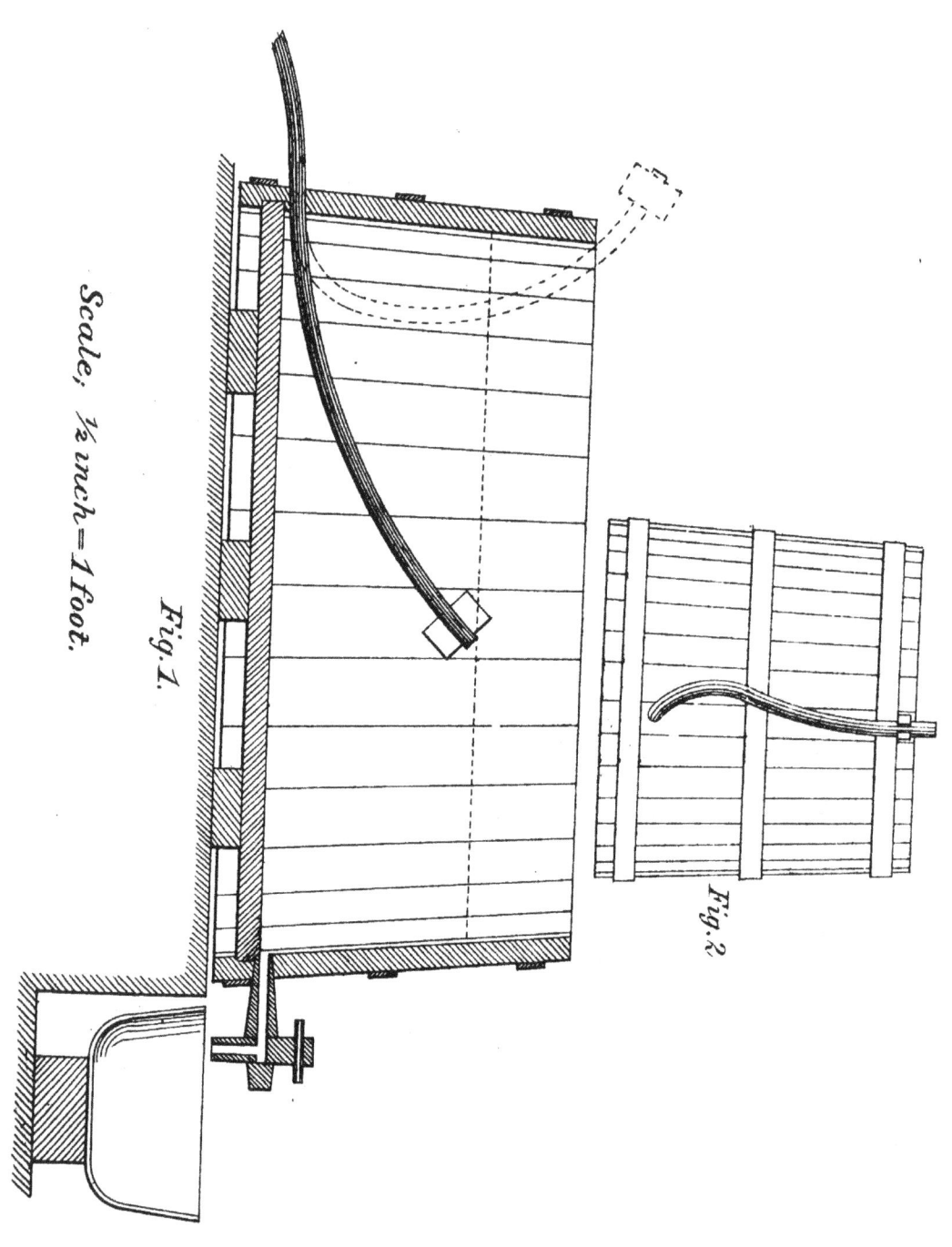

Plate IV.

Scale; ½ inch=1 foot.

Fig. 1

Fig. 2

Plate V.

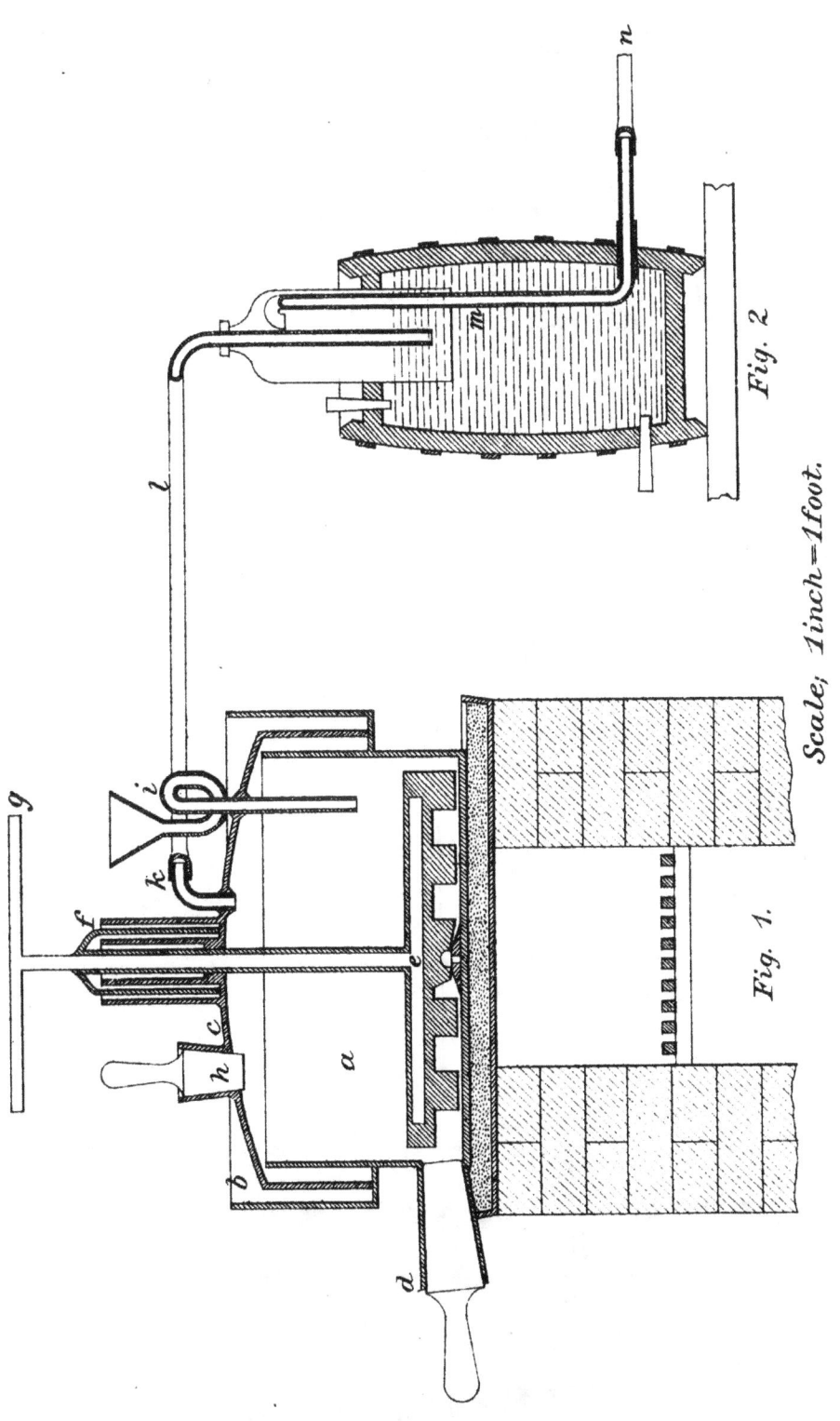

Fig. 2.

Fig. 1.

Scale; 1inch=1foot.

Plate VI.

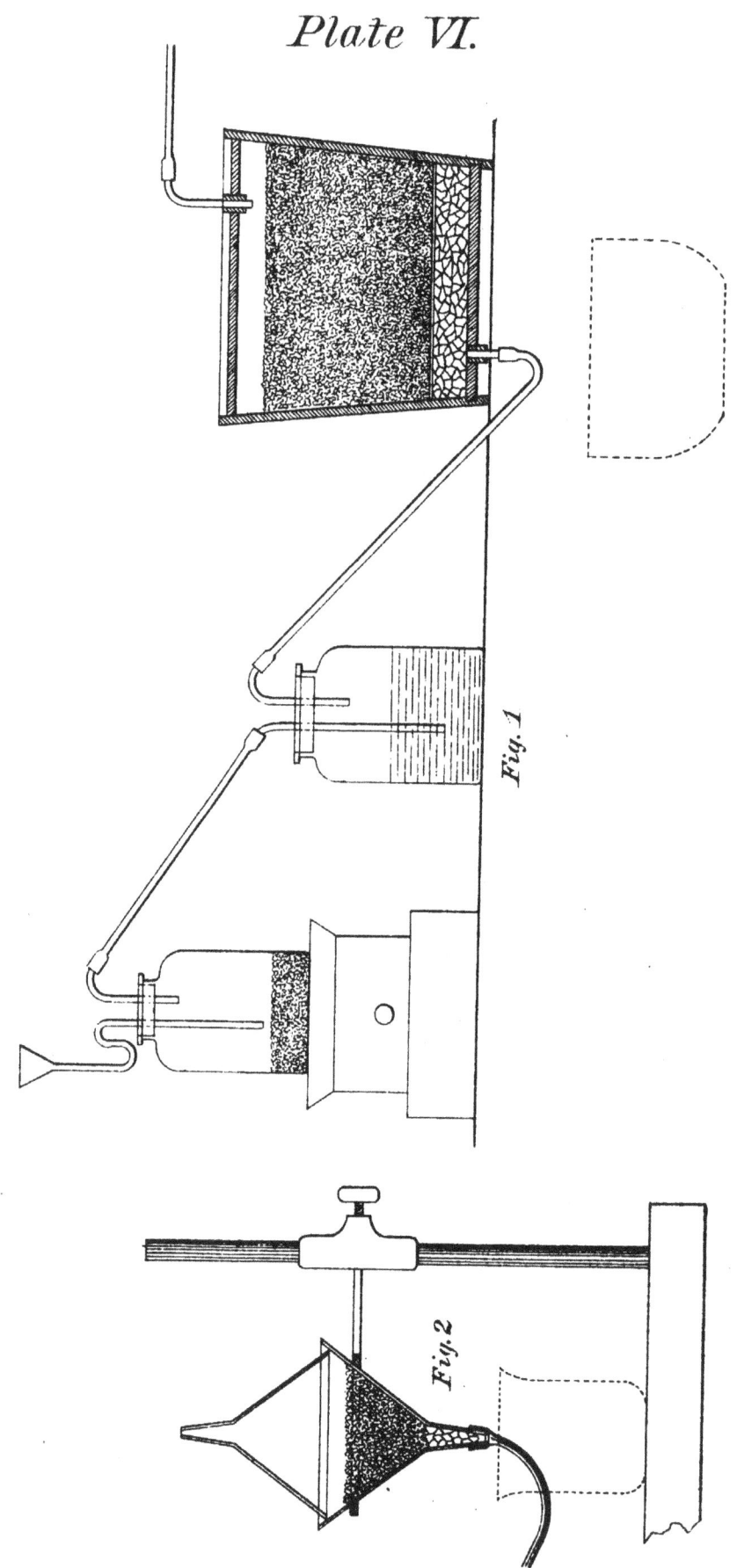

Fig. 1

Fig. 2

Plate VII.

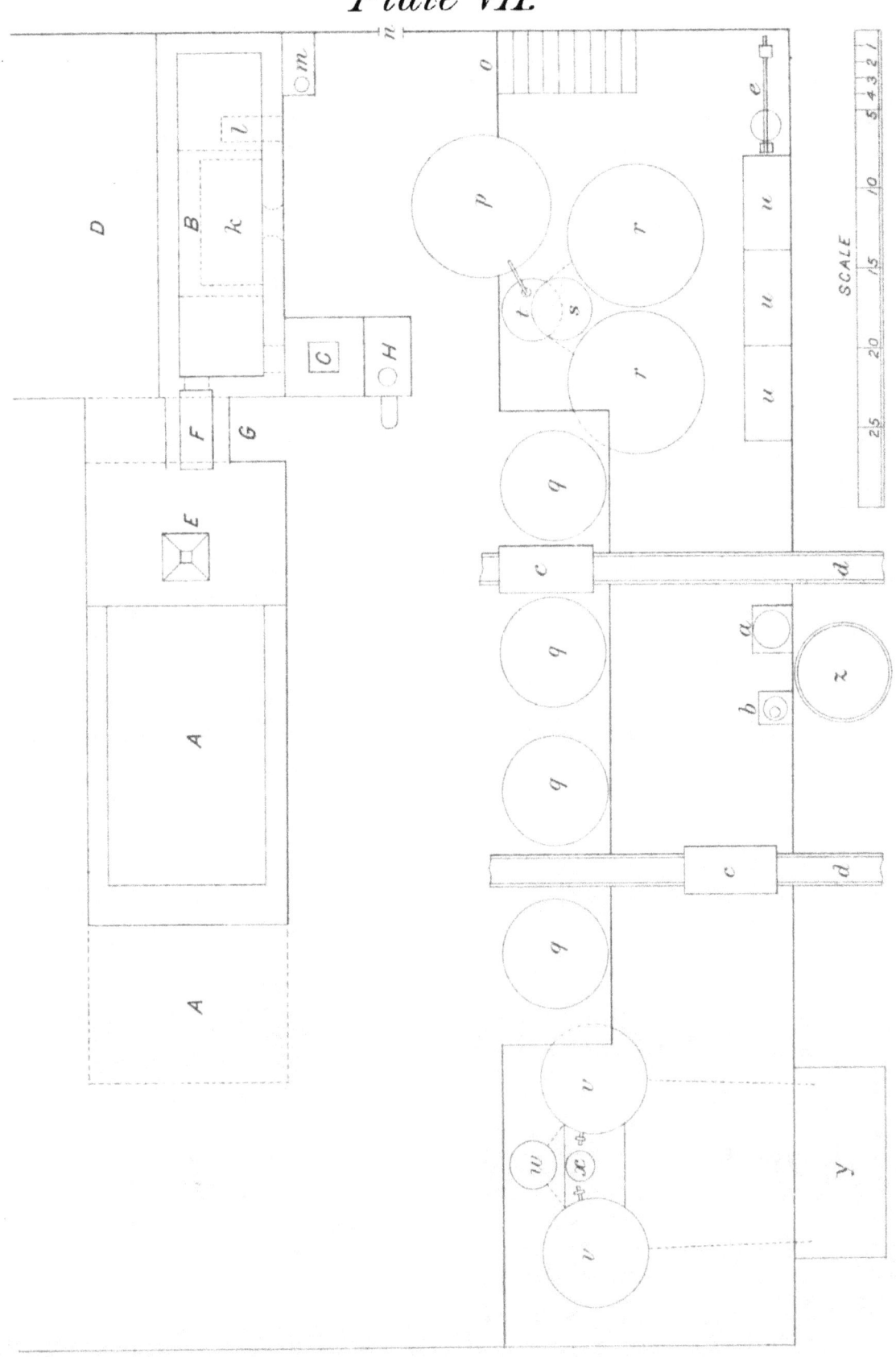

Plan of Works.

Plate VIII.

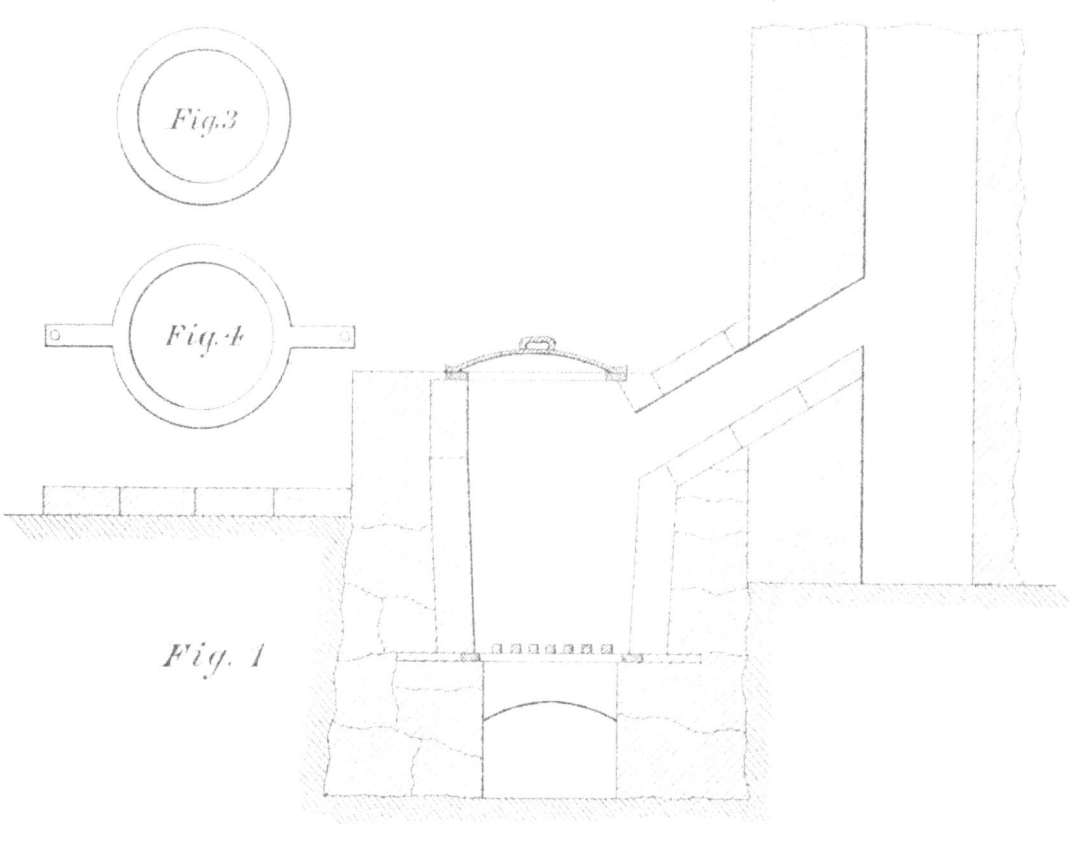

Fig.3

Fig.4

Fig. 1

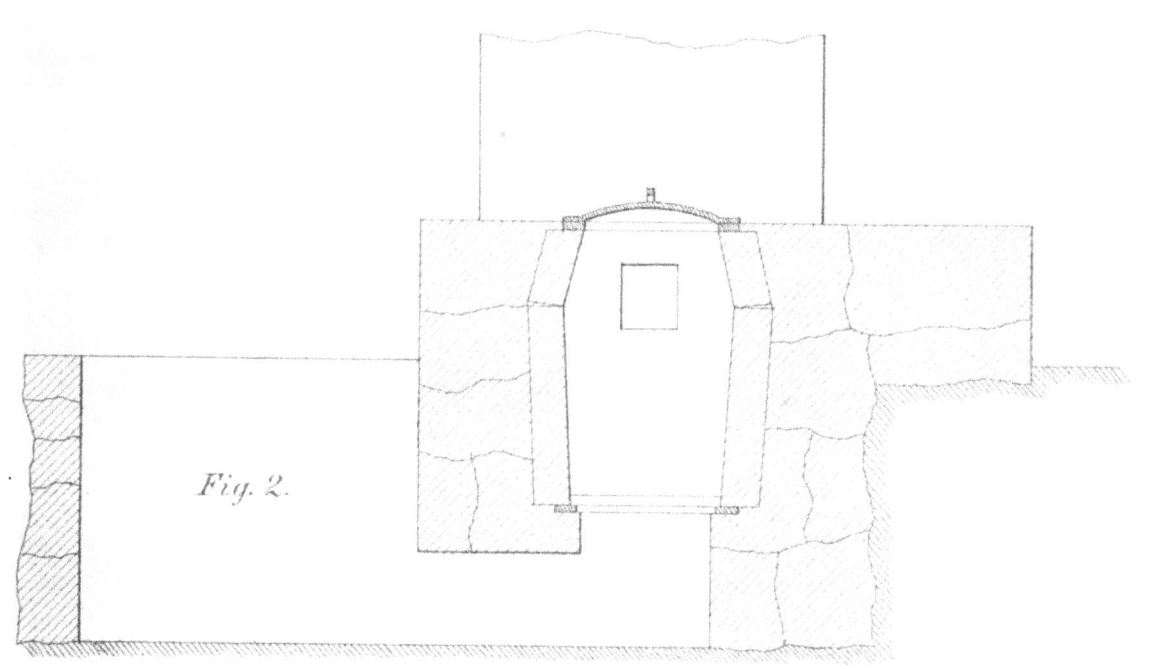

Fig. 2.

Scale; ½ inch = 1 foot.

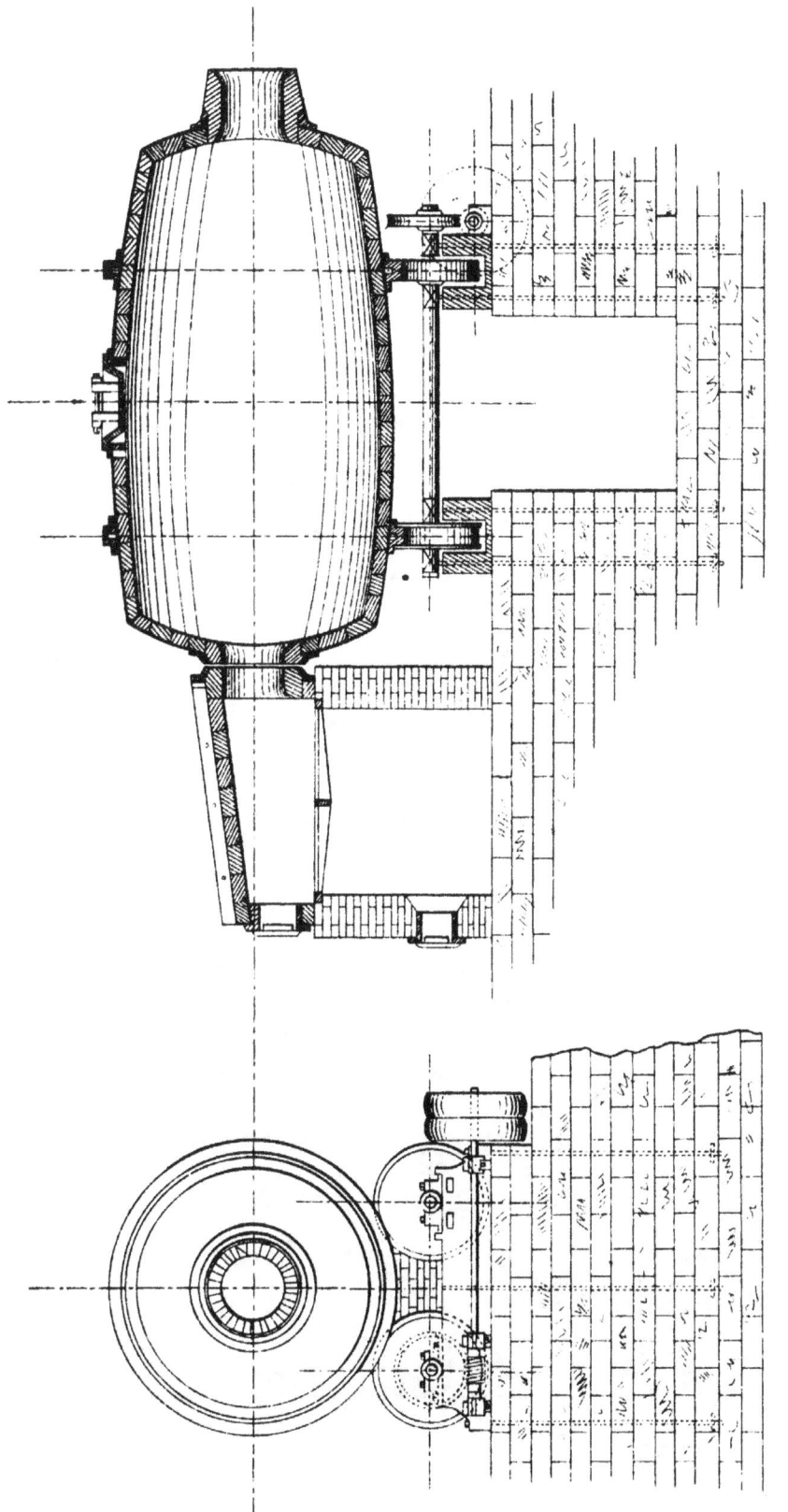

THE BRUNTON FURNACE.

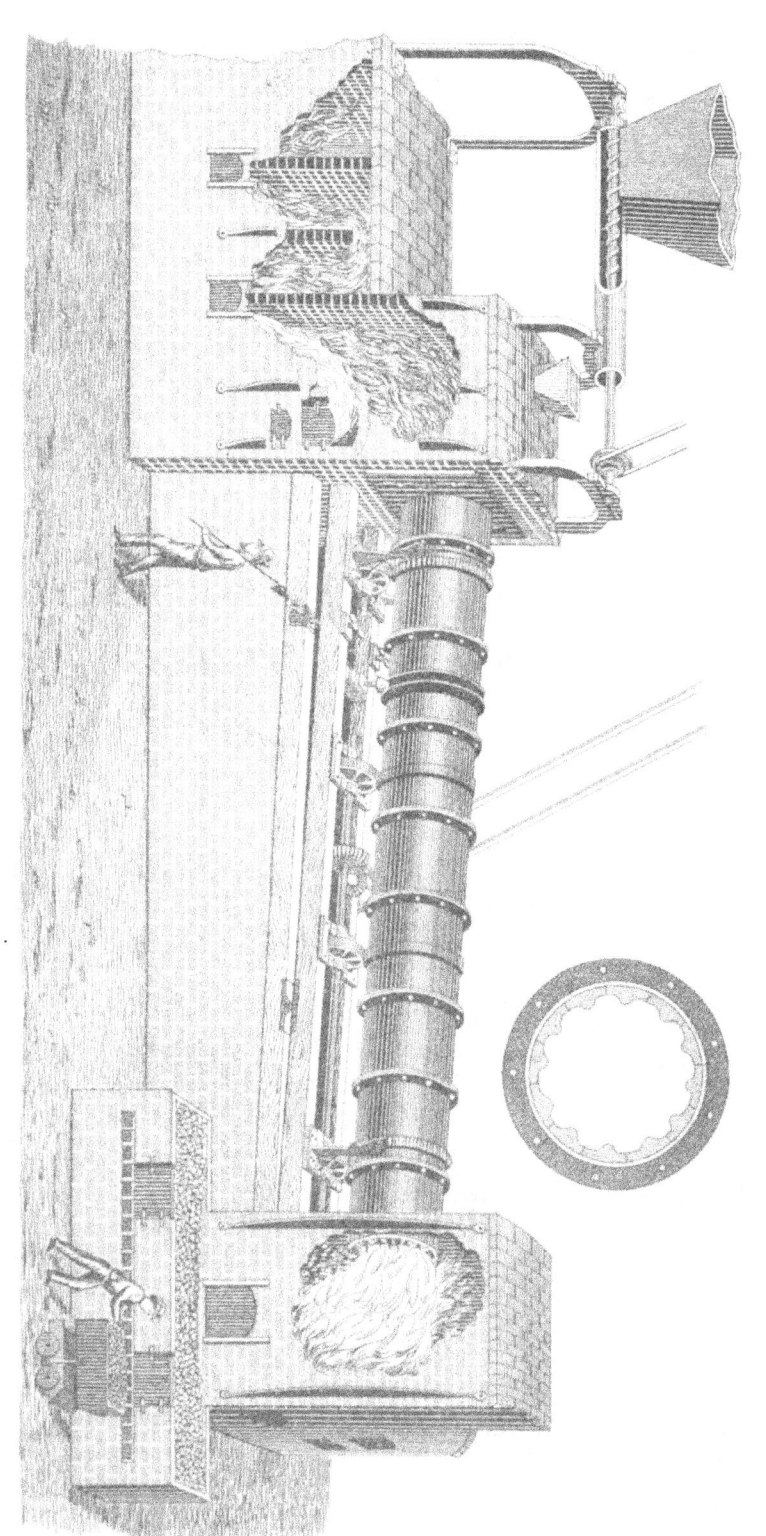

THOMPSON'S IMPROVED HOWELL WHITE FURNACE.

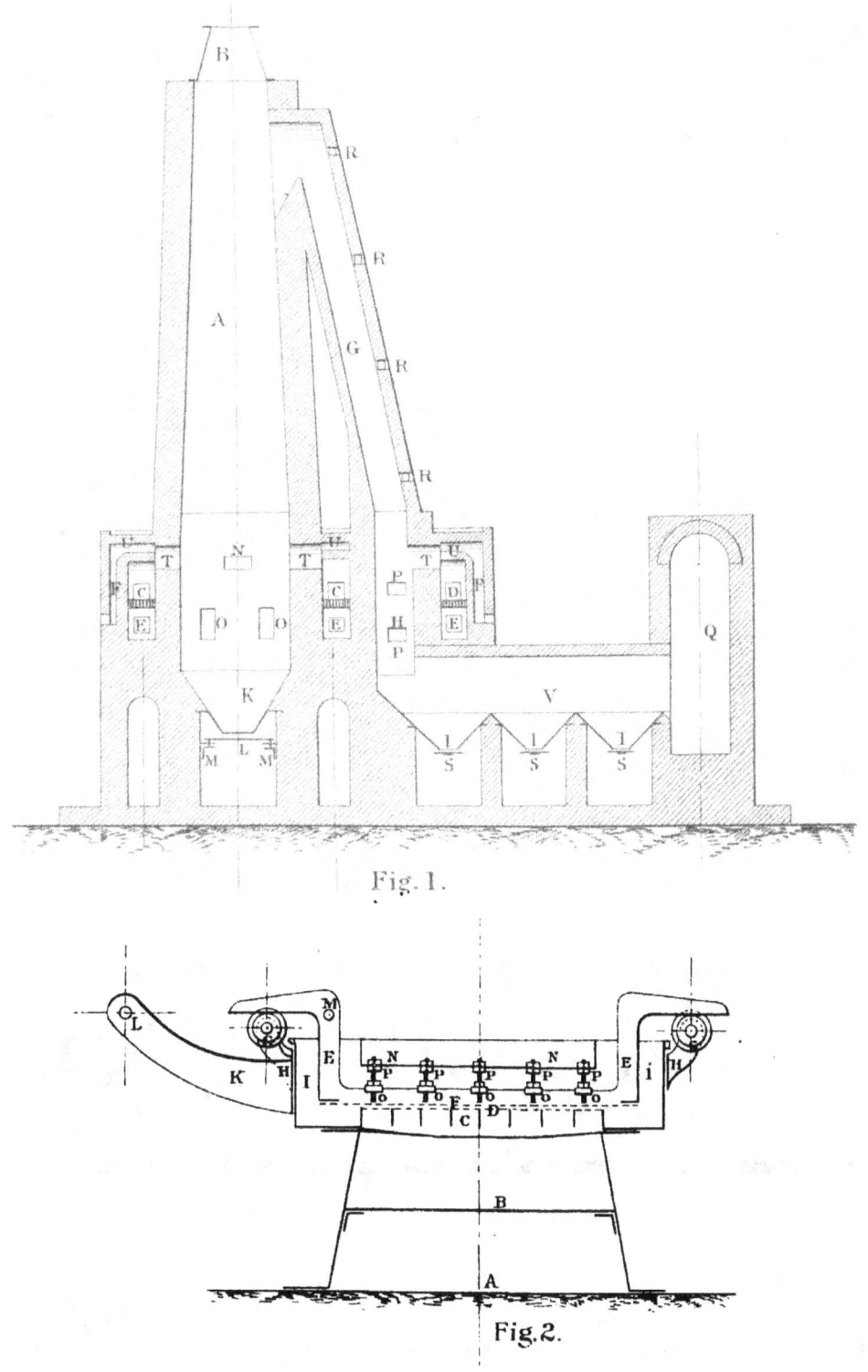

Fig. 1.

Fig. 2.

THE STETEFELDT FURNACE.

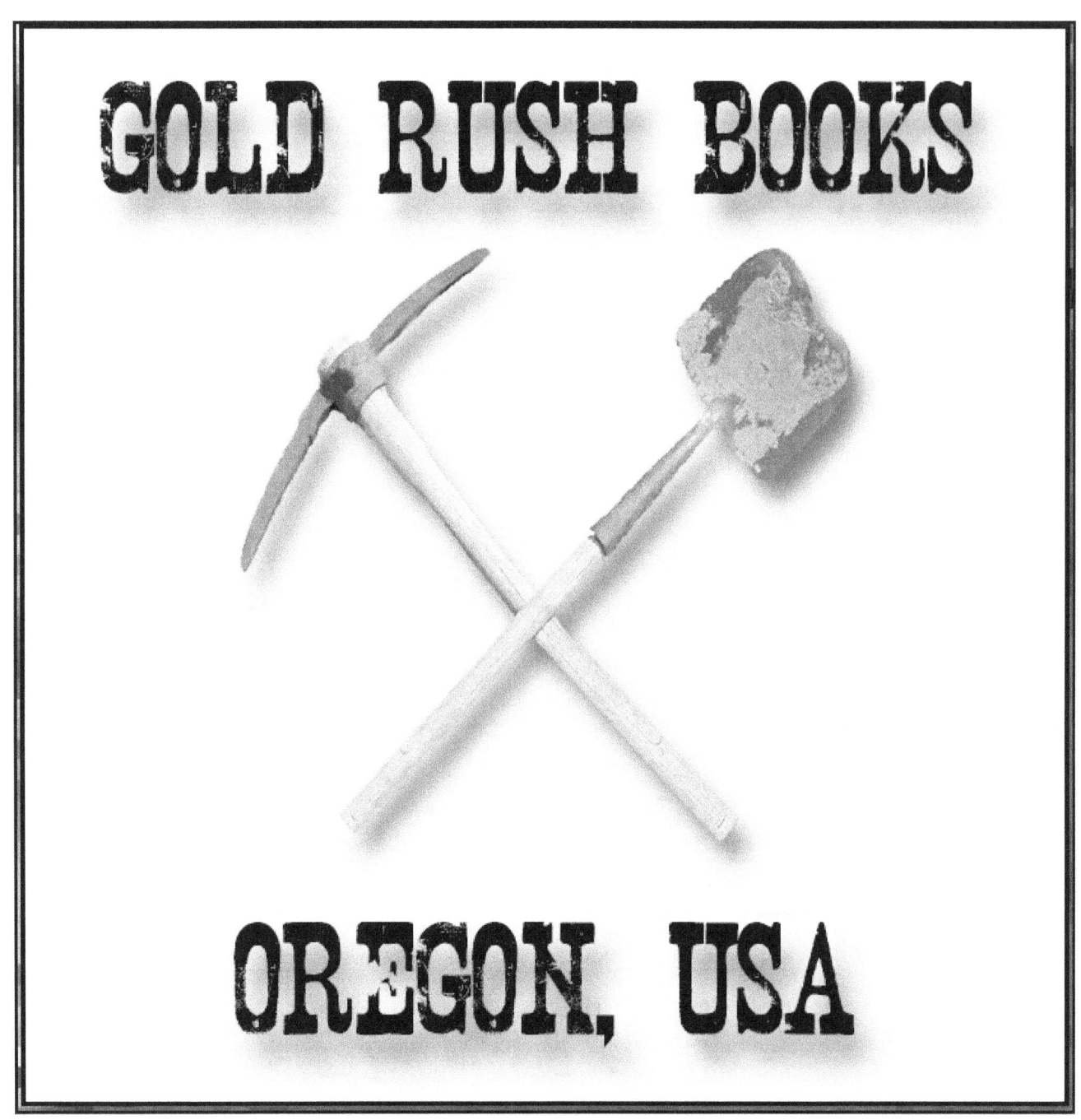

GOLD RUSH BOOKS

OREGON, USA

www.GoldMiningBooks.com

Books On Mining

Visit: www.goldminingbooks.com to order your copies or ask your favorite book seller to offer them.

Mining Books by Kerby Jackson

Gold Dust: Stories From Oregon's Mining Years - Oregon mining historian and prospector, Kerby Jackson, brings you a treasure trove of seventeen stories on Southern Oregon's rich history of gold prospecting, the prospectors and their discoveries, and the breathtaking areas they settled in and made homes. 5" X 8", 98 ppgs. Retail Price: $11.99

The Golden Trail: More Stories From Oregon's Mining Years - In his follow-up to "Gold Dust: Stories of Oregon's Mining Years", this time around, Jackson brings us twelve tales from Oregon's Gold Rush, including the story about the first gold strike on Canyon Creek in Grant County, about the old timers who found gold by the pail full at the Victor Mine near Galice, how Iradel Bray discovered a rich ledge of gold on the Coquille River during the height of the Rogue River War, a tale of two elderly miners on the hunt for a lost mine in the Cascade Mountains, details about the discovery of the famous Armstrong Nugget and others. 5" X 8", 70 ppgs. Retail Price: $10.99

Oregon Mining Books

Geology and Mineral Resources of Josephine County, Oregon - Unavailable since the 1970's, this important publication was originally compiled by the Oregon Department of Geology and Mineral Industries and includes important details on the economic geology and mineral resources of this important mining area in South Western Oregon. Included are notes on the history, geology and development of important mines, as well as insights into the mining of gold, copper, nickel, limestone, chromium and other minerals found in large quantities in Josephine County, Oregon. 8.5" X 11", 54 ppgs. Retail Price: $9.99

Mines and Prospects of the Mount Reuben Mining District - Unavailable since 1947, this important publication was originally compiled by geologist Elton Youngberg of the Oregon Department of Geology and Mineral Industries and includes detailed descriptions, histories and the geology of the Mount Reuben Mining District in Josephine County, Oregon. Included are notes on the history, geology, development and assay statistics, as well as underground maps of all the major mines and prospects in the vicinity of this much neglected mining district. 8.5" X 11", 48 ppgs. Retail Price: $9.99

The Granite Mining District - Notes on the history, geology and development of important mines in the well known Granite Mining District which is located in Grant County, Oregon. Some of the mines discussed include the Ajax, Blue Ribbon, Buffalo, Continental, Cougar-Independence, Magnolia, New York, Standard and the Tillicum. Also included are many rare maps pertaining to the mines in the area. 8.5" X 11", 48 ppgs. Retail Price: $9.99

Ore Deposits of the Takilma and Waldo Mining Districts of Josephine County, Oregon - The Waldo and Takilma mining districts are most notable for the fact that the earliest large scale mining of placer gold and copper in Oregon took place in these two areas. Included are details about some of the earliest large gold mines in the state such as the Llano de Oro, High Gravel, Cameron, Platerica, Deep Gravel and others, as well as copper mines such as the famous Queen of Bronze mine, the Waldo, Lily and Cowboy mines. This volume also includes six maps and 20 original illustrations. 8.5" X 11", 74 ppgs. Retail Price: $9.99

Metal Mines of Douglas, Coos and Curry Counties, Oregon - Oregon mining historian Kerby Jackson introduces us to a classic work on Oregon's mining history in this important re-issue of Bulletin 14C Volume 1, otherwise known as the Douglas, Coos & Curry Counties, Oregon Metal Mines Handbook. Unavailable since 1940, this important publication was originally compiled by the Oregon Department of Geology and Mineral Industries includes detailed descriptions, histories and the geology of over 250 metallic mineral mines and prospects in this rugged area of South West Oregon. 8.5" X 11", 158 ppgs. Retail Price: $19.99

Metal Mines of Jackson County, Oregon - Unavailable since 1943, this important publication was originally compiled by the Oregon Department of Geology and Mineral Industries includes detailed descriptions, histories and the geology of over 450 metallic mineral mines and prospects in Jackson County, Oregon. Included are such famous gold mining areas as Gold Hill, Jacksonville, Sterling and the Upper Applegate. 8.5″ X 11″, 220 ppgs. **Retail Price: $24.99**

Metal Mines of Josephine County, Oregon - Oregon mining historian Kerby Jackson introduces us to a classic work on Oregon's mining history in this important re-issue of Bulletin 14C, otherwise known as the Josephine County, Oregon Metal Mines Handbook. Unavailable since 1952, this important publication was originally compiled by the Oregon Department of Geology and Mineral Industries includes detailed descriptions, histories and the geology of over 500 metallic mineral mines and prospects in Josephine County, Oregon. 8.5″ X 11″, 250 ppgs. **Retail Price: $24.99**

Metal Mines of North East Oregon - Oregon mining historian Kerby Jackson introduces us to a classic work on Oregon's mining history in this important re-issue of Bulletin 14A and 14B, otherwise known as the North East Oregon Metal Mines Handbook. Unavailable since 1941, this important publication was originally compiled by the Oregon Department of Geology and Mineral Industries and includes detailed descriptions, histories and the geology of over 750 metallic mineral mines and prospects in North Eastern Oregon. 8.5″ X 11″, 310 ppgs. **Retail Price: $29.99**

Metal Mines of North West Oregon - Oregon mining historian Kerby Jackson introduces us to a classic work on Oregon's mining history in this important re-issue of Bulletin 14D, otherwise known as the North West Oregon Metal Mines Handbook. Unavailable since 1951, this important publication was originally compiled by the Oregon Department of Geology and Mineral Industries and includes detailed descriptions, histories and the geology of over 250 metallic mineral mines and prospects in North Western Oregon. 8.5″ X 11″, 182 ppgs. **Retail Price: $19.99**

Mines and Prospects of Oregon - Mining historian Kerby Jackson introduces us to a classic mining work by the Oregon Bureau of Mines in this important re-issue of The Handbook of Mines and Prospects of Oregon. Unavailable since 1916, this publication includes important insights into hundreds of gold, silver, copper, coal, limestone and other mines that operated in the State of Oregon around the turn of the 19th Century. Included are not only geological details on early mines throughout Oregon, but also insights into their history, production, locations and in some cases, also included are rare maps of their underground workings. 8.5″ X 11″, 314 ppgs. **Retail Price: $24.99**

Lode Gold of the Klamath Mountains of Northern California and South West Oregon (See California Mining Books)

Mineral Resources of South West Oregon - Unavailable since 1914, this publication includes important insights into dozens of mines that once operated in South West Oregon, including the famous gold fields of Josephine and Jackson Counties, as well as the Coal Mines of Coos County. Included are not only geological details on early mines throughout South West Oregon, but also insights into their history, production and locations. 8.5″ X 11″, 154 ppgs. **Retail Price: $11.99**

Chromite Mining in The Klamath Mountains of California and Oregon (See California Mining Books)

Southern Oregon Mineral Wealth - Unavailable since 1904, this rare publication provides a unique snapshot into the mines that were operating in the area at the time. Included are not only geological details on early mines throughout South West Oregon, but also insights into their history, production and locations. Some of the mining areas include Grave Creek, Greenback, Wolf Creek, Jump Off Joe Creek, Granite Hill, Galice, Mount Reuben, Gold Hill, Galls Creek, Kane Creek, Sardine Creek, Birdseye Creek, Evans Creek, Foots Creek, Jacksonville, Ashland, the Applegate River, Waldo, Kerby and the Illinois River, Althouse and Sucker Creek, as well as insights into local copper mining and other topics. 8.5″ X 11″, 64 ppgs. **Retail Price: $8.99**

Geology and Ore Deposits of the Takilma and Waldo Mining Districts - Unavailable since the 1933, this publication was originally compiled by the United States Geological Survey and includes details on gold and copper mining in the Takilma and Waldo Districts of Josephine County, Oregon. The Waldo and Takilma mining districts are most notable for the fact that the earliest large scale mining of placer gold and copper in Oregon took place in these two areas. Included in this report are details about some of the earliest large gold mines in the state such as the Llano de Oro, High Gravel, Cameron, Platerica, Deep Gravel and others, as well as copper mines such as the famous Queen of Bronze mine, the Waldo, Lily and Cowboy mines. In addition to geological examinations, insights are also provided into the production, day to day operations and early histories of these mines, as well as calculations of known mineral reserves in the area. This volume also includes six maps and 20 original illustrations. 8.5″ X 11″, 74 ppgs. **Retail Price: $9.99**

Gold Mines of Oregon - Oregon mining historian Kerby Jackson introduces us to a classic work on Oregon's mining history in this important re-issue of Bulletin 61, otherwise known as "Gold and Silver In Oregon". Unavailable since 1968, this important publication was originally compiled by geologists Howard C. Brooks and Len Ramp of the Oregon Department of Geology and Mineral Industries and includes detailed descriptions, histories and the geology of over 450 gold mines Oregon. Included are notes on the history, geology and gold production statistics of all the major mining areas in Oregon including the Klamath Mountains, the Blue Mountains and the North Cascades. While gold is where you find it, as every miner knows, the path to success is to prospect for gold where it was previously found. **8.5" X 11", 344 ppgs. Retail Price: $24.99**

Mines and Mineral Resources of Curry County Oregon - Originally published in 1916, this important publication on Oregon Mining has not been available for nearly a century. Included are rare insights into the history, production and locations of dozens of gold mines in Curry County, Oregon, as well as detailed information on important Oregon mining districts in that area such as those at Agness, Bald Face Creek, Mule Creek, Boulder Creek, China Diggings, Collier Creek, Elk River, Gold Beach, Rock Creek, Sixes River and elsewhere. Particular attention is especially paid to the famous beach gold deposits of this portion of the Oregon Coast. **8.5" X 11", 140 ppgs. Retail Price: $11.99**

Chromite Mining in South West Oregon - Originally published in 1961, this important publication on Oregon Mining has not been available for nearly a century. Included are rare insights into the history, production and locations of nearly 300 chromite mines in South Western Oregon. **8.5" X 11", 184 ppgs. Retail Price: $14.99**

Mineral Resources of Douglas County Oregon - Originally published in 1972, this important publication on Oregon Mining has not been available for nearly forty years. Included are rare insights into the geology, history, production and locations of numerous gold mines and other mining properties in Douglas County, Oregon. **8.5" X 11", 124 ppgs. Retail Price: $11.99**

Mineral Resources of Coos County Oregon - Originally published in 1972, this important publication on Oregon Mining has not been available for nearly forty years. Included are rare insights into the geology, history, production and locations of numerous gold mines and other mining properties in Coos County, Oregon. **8.5" X 11", 100 ppgs. Retail Price: $11.99**

Mineral Resources of Lane County Oregon - Originally published in 1938, this important publication on Oregon Mining has not been available for nearly seventy five years. Included are extremely rare insights into the geology and mines of Lane County, Oregon, in particular in the Bohemia, Blue River, Oakridge, Black Butte and Winberry Mining Districts. **8.5" X 11", 82 ppgs. Retail Price: $9.99**

Mineral Resources of the Upper Chetco River of Oregon: Including the Kalmiopsis Wilderness - Originally published in 1975, this important publication on Oregon Mining has not been available for nearly forty years. Withdrawn under the 1872 Mining Act since 1984, real insight into the minerals resources and mines of the Upper Chetco River has long been unavailable due to the remoteness of the area. Despite this, the decades of battle between property owners and environmental extremists over the last private mining inholding in the area has continued to pique the interest of those interested in mining and other forms of natural resource use. Gold mining began in the area in the 1850's and has a rich history in this geographic area, even if the facts surrounding it are little known. Included are twenty two rare photographs, as well as insights into the Becca and Morning Mine, the Emmly Mine (also known as Emily Camp), the Frazier Mine, the Golden Dream or Higgins Mine, Hustis Mine, Peck Mine and others. **8.5" X 11", 64 ppgs. Retail Price: $8.99**

Gold Dredging in Oregon - Originally published in 1939, this important publication on Oregon Mining has not been available for nearly seventy five years. Included are extremely rare insights into the history and day to day operations of the dragline and bucketline gold dredges that once worked the placer gold fields of South West and North East Oregon in decades gone by. Also included are details into the areas that were worked by gold dredges in Josephine, Jackson, Baker and Grant counties, as well as the economic factors that impacted this mining method. This volume also offers a unique look into the values of river bottom land in relation to both farming and mining, in how farm lands were mined, re-soiled and reclamated after the dredges worked them. Featured are hard to find maps of the gold dredge fields, as well as rare photographs from a bygone era. **8.5" X 11", 86 ppgs. Retail Price: $8.99**

Quick Silver Mining in Oregon - Originally published in 1963, this important publication on Oregon Mining has not been available for over fifty years. This publication includes details into the history and production of Elemental Mercury or Quicksilver in the State of Oregon. **8.5" X 11", 238 ppgs. Retail Price: $15.99**

Mines of the Greenhorn Mining District of Grant County Oregon - Originally published in 1948, this important publication on Oregon Mining has not been available for over sixty five years. In this publication are rare insights into the mines of the famous Greenhorn Mining District of Grant County, Oregon, especially the famous Morning Mine. Also included are details on the Tempest, Tiger, Bi-Metallic, Windsor, Psyche, Big Johnny, Snow Creek, Banzette and Paramount Mines, as well as prospects in the vicinities in the famous mining areas of Mormon Basin, Vinegar Basin and Desolation Creek. Included are hard to find mine maps and dozens of rare photographs from the bygone era of Grant County's rich mining history. **8.5" X 11", 72 ppgs. Retail Price: $9.99**

Geology of the Wallowa Mountains of Oregon: Part I (Volume 1) - Originally published in 1938, this important publication on Oregon Mining has not been available for nearly seventy five years. Included are details on the geology of this unique portion of North Eastern Oregon. This is the first part of a two book series on the area. Accompanying the text are rare photographs and historic maps.8.5" X 11", 92 ppgs. Retail Price: $9.99

Geology of the Wallowa Mountains of Oregon: Part II (Volume 2) - Originally published in 1938, this important publication on Oregon Mining has not been available for nearly seventy five years. Included are details on the geology of this unique portion of North Eastern Oregon. This is the first part of a two book series on the area. Accompanying the text are rare photographs and historic maps.8.5" X 11", 94 ppgs. Retail Price: $9.99

Field Identification of Minerals For Oregon Prospectors - Originally published in 1940, this important publication on Oregon Mining has not been available for nearly seventy five years. Included in this volume is an easy system for testing and identifying a wide range of minerals that might be found by prospectors, geologists and rockhounds in the State of Oregon, as well as in other locales. Topics include how to put together your own field testing kit and how to conduct rudimentary tests in the field. This volume is written in a clear and concise way to make it useful even for beginners. 8.5" X 11", 158 ppgs. Retail Price: $14.99

The Bohemia Mining District of Oregon - Originally published in 1900, this important publication on Oregon Mining has not been available for over a century. Included in this volume are important insights into the famous Bohemia Mining District of Oregon, including the histories and locations of important gold mines in the area such as the Ophir Mine, Clarence, Acturas, Peek-a-boo, White Swan, Combination Mine, the Musick Mine, The California, White Ghost, The Mystery, Wall Street, Vesuvius, Story, Lizzie Bullock, Delta, Elsie Dora, Golden Slipper, Broadway, Champion Mine, Knott, Noonday, Helena, White Wings, Riverside and others. Also included are notes on the nearby Blue River Mining District. 8.5" X 11", 58 ppgs. Retail Price: $9.99

The Gold Fields of Eastern Oregon - Unavailable since 1900, this publication was originally compiled by the Baker City Chamber of Commerce Offering important insights into the gold mining history of Eastern Oregon, "The Gold Fields of Eastern Oregon" sheds a rare light on many of the gold mines that were operating at the turn of the 19th Century in Baker County and Grant County in North Eastern Oregon. Some of the areas featured include the Cable Cove District, Baisely-Elhorn, Granite, Red Boy, Bonanza, Susanville, Sparta, Virtue, Vaughn, Sumpter, Burnt River, Rye Valley and other mining districts. Included is basic information on not only many gold mines that are well known to those interested in Eastern Oregon mining history, but also many mines and prospects which have been mostly lost to the passage of time. Accompanying are numerous rare photos **8.5" X 11", 78 ppgs. Retail Price: $10.99**

Gold Mining in Eastern Oregon - Originally published in 1938, this important publication on Oregon Mining has not been available for over a century. Included in this volume are important insights into the famous mining districts of Eastern Oregon during the late 1930's. Particular attention is given to those gold mines with milling and concentrating facilities in the Greenhorn, Red Boy, Alamo, Bonanza, Granite, Cable Cove, Cracker Creek, Virtue, Keating, Medical Springs, Sanger, Sparta, Chicken Creek, Mormon Basin, Connor Creek, Cornucopia and the Bull Run Mining Districts. Some of the mines featured include the Ben Harrison, North Pole-Columbia, Highland Maxwell, Baisley-Elkhorn, White Swan, Balm Creek, Twin Baby, Gem of Sparta, New Deal, Gleason, Gifford-Johnson, Cornucopia, Record, Bull Run, Orion and others. Of particular interest are the mill flow sheets and descriptions of milling operations of these mines. 8.5" X 11", 68 ppgs. Retail Price: $8.99

The Gold Belt of the Blue Mountains of Oregon - Originally published in 1901, this important publication on Oregon Mining has not been available for over a century. Included in this volume are rare insights into the gold deposits of the Blue Mountains of North East Oregon, including the history of their early discovery and early production. Extensive details are offered on this important mining area's mineralogy and economic geology, as well as insights into nearby gold placers, silver deposits and copper deposits. Featured are the Elkhorn and Rock Creek mining districts, the Pocahontas district, Auburn and Minersville districts, Sumpter and Cracker Creek, Cable Cove, the Camp Carson district, Granite, Alamo, Greenhorn, Robinsonville, the Upper Burnt River Valley and Bonanza districts, Susanville, Quartzburg, Canyon Creek, Virtue, the Copper Butte district, the North Powder River, Sparta, Eagle Creek, Cornucopia, Pine Creek, Lower Powder River, the Upper Snake River Canyon, Rye Valley, Lower Burnt River Valley, Mormon Basin, the Malheur and Clarks Creek districts, Sutton Creek and others. Of particular interest are important details on numerous gold mines and prospects in these mining districts, including their locations, histories, geology and other important information, as well as information on silver, copper and fire opal deposits. 8.5" X 11", 250 ppgs. Retail Price: $24.99

<u>Mining in the Cascades Range of Oregon</u> - Originally published in 1938, this important publication on Oregon Mining has not been available for over seventy five years. Included in this volume are rare insights into the gold mines and other types of metal mines in the Cascades Mountain Range of Oregon. Some of the important mining areas covered include the famous Bohemia Mining District, the North Santiam Mining District, Quartzville Mining District, Blue River Mining District, Fall Creek Mining District, Oakridge District, Zinc District, Buzzard-Al Sarena District, Grand Cove, Climax District and Barron Mining District. Of particular interest are important details on over 100 mines and prospects in these mining districts, including their locations, histories, geology and other important information. **8.5" X 11", 170 ppgs. Retail Price: $14.99**

<u>Beach Gold Placers of the Oregon Coast</u> - Originally published in 1934, this important publication on Oregon Mining has not been available for over 80 years. Included in this volume are rare insights into the beach gold deposits of the State of Oregon, including their locations, occurance, composition and geology. Of particular interest is information on placer platinum in Oregon's rich beach deposits. Also included are the locations and other information on some famous Oregon beach mines, including the Pioneer, Eagle, Chickamin, Iowa and beach placer mines north of the mouth of the Rogue River. **8.5" X 11", 60 ppgs. Retail Price: $8.99**

Idaho Mining Books

<u>Gold in Idaho</u> - Unavailable since the 1940's, this publication was originally compiled by the Idaho Bureau of Mines and includes details on gold mining in Idaho. Included is not only raw data on gold production in Idaho, but also valuable insight into where gold may be found in Idaho, as well as practical information on the gold bearing rocks and other geological features that will assist those looking for placer and lode gold in the State of Idaho. This volume also includes thirteen gold maps that greatly enhance the practical usability of the information contained in this small book detailing where to find gold in Idaho. **8.5" X 11", 72 ppgs. Retail Price: $9.99**

<u>Geology of the Couer D'Alene Mining District of Idaho</u> - Unavailable since 1961, this publication was originally compiled by the Idaho Bureau of Mines and Geology and includes details on the mining of gold, silver and other minerals in the famous Coeur D'Alene Mining District in Northern Idaho. Included are details on the early history of the Coeur D'Alene Mining District, local tectonic settings, ore deposit features, information on the mineral belts of the Osburn Fault, as well as detailed information on the famous Bunker Hill Mine, the Dayrock Mine, Galena Mine, Lucky Friday Mine and the infamous Sunshine Mine. This volume also includes sixteen hard to find maps. **8.5" X 11", 70 ppgs. Retail Price: $9.99**

<u>The Gold Camps and Silver Cities of Idaho</u> - Originally published in 1963, this important publication on Idaho Mining has not been available for nearly fifty years. Included are rare insights into the history of Idaho's Gold Rush, as well as the mad craze for silver in the Idaho Panhandle. Documented in fine detail are the early mining excitements at Boise Basin, at South Boise, in the Owyhees, at Deadwood, Long Valley, Stanley Basin and Robinson Bar, at Atlanta, on the famous Boise River, Volcano, Little Smokey, Banner, Boise Ridge, Hailey, Leesburg, Lemhi, Pearl, at South Mountain, Shoup and Ulysses, Yellow Jacket and Loon Creek. The story follows with the appearance of Chinese miners at the new mining camps on the Snake River, Black Pine, Yankee Fork, Bay Horse, Clayton, Heath, Seven Devils, Gibbonsville, Vienna and Sawtooth City. Also included are special sections on the Idaho Lead and Silver mines of the late 1800's, as well as the mining discoveries of the early 1900's that paved the way for Idaho's modern mining and mineral industry. Lavishly illustrated with rare historic photos, this volume provides a one of a kind documentary into Idaho's mining history that is sure to be enjoyed by not only modern miners and prospectors who still scour the hills in search of nature's treasures, but also those enjoy history and tromping through overgrown ghost towns and long abandoned mining camps. **8.5" X 11", 186 ppgs. Retail Price: $14.99**

<u>Ore Deposits and Mining in North Western Custer County Idaho</u> - Unavailable since 1913, this important publication was originally published by the Us Department of the Interior and has been unavailable for a century. Included are fine details on the geology, geography, gold placers and gold and silver bearing quartz veins of the mining region of North West Custer County, Idaho. Of particular interest is a rare look at the mines and prospects of the region, including those such as the Ramshorn Mine, SkyLark, Riverview, Excelsior, Beardsley, Pacific, Hoosier, Silver Brick, Forest Rose and dozens of others in the Bay Horse Mining District. Also covered are the mines of the Yankee Fork District such as the Lucky Boy, Badger, Black, Enterprise, Charles Dickens, Morrison, Golden Sunbeam, Montana, Golden Gate and others, as well as those in the Loon Mining District. **8.5" X 11", 126 ppgs. Retail Price: $12.99**

Gold Rush To Idaho - Unavailable since 1963, this important publication was originally published by the Idaho Bureau of Mines and has been unavailable for 50 years. "Gold Rush To Idaho" revisits the earliest years of the discovery of gold in Idaho Territory and introduces us to the conditions that the pioneer gold seekers met when they blazed a trail through the wilderness of Idaho's mountains and discovered the precious yellow metal at Oro Fino and Pierce. Subsequent rushes followed at places like Elk City, Newsome, Clearwater Station, Florence, Warrens and elsewhere. Of particular interest is a rare look at the hardships that the first miners in Idaho met with during their day to day existences and their attempts to bring law and order to their mining camps. **8.5" X 11", 88 ppgs. Retail Price: $9.99**

The Geology and Mines of Northern Idaho and North Western Montana - Unavailable since 1909, this important publication was originally published by the Us Department of the Interior and has been unavailable for a century. Included are fine details on the geology and geography of the mining regions of Northern Idaho and North Western Montana. Of particular interest is a rare look at the mines and prospects of the region, including those in the Pine Creek Mining District, Lake Pend Oreille district, Troy Mining District, Sylvanite District, Cabinet Mining District, Prospect Mining District and the Missoula Valley. Some of the mines featured include the Iron Mountain, Silver Butte, Snowshoe, Grouse Mountain Mine and others. **8.5" X 11", 142 ppgs. Retail Price: $12.99**

Mining in the Alturas Quadrangle of Blaine County Idaho - Unavailable since 1922, this important publication was originally published by the Idaho Bureau of Mines and has been unavailable for ninety years. Topics include the geology, rock formations and the formation of ore deposits in this important mining area of Idaho. Of particular focus is information on the local geology, quartz veins and ore deposits of this portion of Idaho. Included are hard to find details, including the descriptions and locations of numerous gold and silver mines in the area including the Silver King, Pilgrim, Columbia, Lone Jack, Sunbeam, Pride of the West, Lucky Boy, Scotia, Atlanta, Beaver-Bidwell and others mines and prospects. **8.5" X 11", 56 ppgs. Retail Price: $8.99**

Mining in Lemhi County Idaho - Originally published in 1913, this important book on Idaho Mining has not been available to miners for over a century. Included are rare insights into hundreds of gold, silver, copper and other mines in this famous Idaho mining area. Details include the locations, geology, history, production and other facts of the mines of this region, not only gold and silver hardrock mines, but also gold placer mines, lead-silver deposits, copper mines, cobalt-nickel deposits, tungsten and tin mines . It is lavishly illustrated with hard to find photos of the period and rare mining maps. Some of the vicinities featured include the Nicholia Mining District, Spring Mountain District, Texas District, Blue Wing District, Junction District, McDevitt District, Pratt Creek, Eldorado District, Kirtley Creek, Carmen Creek, Gibbonsville, Indian Creek, Mineral Hill District, Mackinaw, Eureka District, Blackbird District, YellowJacket District, Gravel Range District, Junction District, Parker Mountain and other mining districts. **8.5" X 11", 226 ppgs. Retail Price: $19.99**

Utah Mining Books

Fluorite in Utah - Unavailable since 1954, this publication was originally compiled by the USGS, State of Utah and U.S. Atomic Energy Commission and details the mining of fluorspar, also known as fluorite in the State of Utah. Included are details on the geology and history of fluorspar (fluorite) mining in Utah, including details on where this unique gem mineral may be found in the State of Utah. **8.5" X 11", 60 ppgs. Retail Price: $8.99**

California Mining Books

The Tertiary Gravels of the Sierra Nevada of California - Mining historian Kerby Jackson introduces us to a classic mining work by Waldemar Lindgren in this important re-issue of The Tertiary Gravels of the Sierra Nevada of California. Unavailable since 1911, this publication includes details on the gold bearing ancient river channels of the famous Sierra Nevada region of California. **8.5" X 11", 282 ppgs. Retail Price: $19.99**

The Mother Lode Mining Region of California - Unavailable since 1900, this publication includes details on the gold mines of California's famous Mother Lode gold mining area. Included are details on the geology, history and important gold mines of the region, as well as insights into historic mining methods, mine timbering, mining machinery, mining bell signals and other details on how these mines operated. Also included are insights into the gold mines of the California Mother Lode that were in operation during the first sixty years of California's mining history. **8.5" X 11", 176 ppgs. Retail Price: $14.99**

Lode Gold of the Klamath Mountains of Northern California and South West Oregon - Unavailable since 1971, this publication was originally compiled by Preston E. Hotz and includes details on the lode mining districts of Oregon and California's Klamath Mountains. Included are details on the geology, history and important lode mines of the French Gulch, Deadwood, Whiskeytown, Shasta, Redding, Muletown, South Fork, Old Diggings, Dog Creek (Delta), Bully Choop (Indian Creek), Harrison Gulch, Hayfork, Minersville, Trinity Center, Canyon Creek, East Fork, New River, Denny, Liberty (Black Bear), Cecilville, Callahan, Yreka, Fort Jones and Happy Camp mining districts in California, as well as the Ashland, Rogue River, Applegate, Illinois River, Takilma, Greenback, Galice, Silver Peak, Myrtle Creek and Mule Creek districts of South Western Oregon. Also included are insights into the mineralization and other characteristics of this important mining region. **8.5" X 11", 100 ppgs. Retail Price: $10.99**

Mines and Mineral Resources of Shasta County, Siskiyou County, Trinity County: California - Unavailable since 1915, this publication was originally compiled by the California State Mining Bureau and includes details on the gold mines of this area of Northern California. Also included are insights into the mineralization and other characteristics of this important mining region, as well as the location of historic gold mines. **8.5" X 11", 204 ppgs. Retail Price: $19.99**

Geology of the Yreka Quadrangle, Siskiyou County, California - Unavailable since 1977, this publication was originally compiled by Preston E. Hotz and includes details on the geology of the Yreka Quadrangle of Siskiyou County, California. Also included are insights into the mineralization and other characteristics of this important mining region. **8.5" X 11", 78 ppgs. Retail Price: $7.99**

Mines of San Diego and Imperial Counties, California - Originally published in 1914, this important publication on California Mining has not been available for a century. This publication includes important information on the early gold mines of San Diego and Imperial County, which were some of the first gold fields mined in California by early Spanish and Mexican miners before the 49ers came on the scene. Included are not only details on early mining methods in the area, production statistics and geological information, but also the location of the early gold mines that helped make California "The Golden State". Also included are details on the mining of other minerals such as silver, lead, zinc, manganese, tungsten, vanadium, asbestos, barite, borax, cement, clay, dolomite, fluospar, gem stones, graphite, marble, salines, petroleum, stronium, talc and others. **8.5" X 11", 116 ppgs. Retail Price: $12.99**

Mines of Sierra County, California - Unavailable since 1920, this publication was originally compiled by the California State Mining Bureau and includes details on the gold mines of Sierra County, California. Also included are insights into the mineralization and other characteristics of this important mining region, as well as the location of historic gold mines. **8.5" X 11", 156 ppgs. Retail Price: $19.99**

Mines of Plumas County, California - Unavailable since 1918, this publication was originally compiled by the California State Mining Bureau and includes details on the gold mines of Plumas County, California. Also included are insights into the mineralization and other characteristics of this important mining region, as well as the location of historic gold mines. **8.5" X 11", 200 ppgs. Retail Price: $19.99**

Mines of El Dorado, Placer, Sacramento and Yuba Counties, California - Originally published in 1917, this important publication on California Mining has not been available for nearly a century. This publication includes important information on the early gold mines of El Dorado County, Placer County, Sacramento County and Yuba County, which were some of the first gold fields mined by the Forty-Niners during the California Gold Rush. Included are not only details on early mining methods in the area, production statistics and geological information, but also the location of the early gold mines that helped make California "The Golden State". Also included are insights into the early mining of chrome, copper and other minerals in this important mining area. **8.5" X 11", 204 ppgs. Retail Price: $19.99**

Mines of Los Angeles, Orange and Riverside Counties, California - Originally published in 1917, this important publication on California Mining has not been available for nearly a century. This publication includes important information on the early gold mines of Los Angeles County, Orange County and Riverside County, which were some of the first gold fields mined in California by early Spanish and Mexican miners before the 49ers came on the scene. Included are not only details on early mining methods in the area, production statistics and geological information, but also the location of the early gold mines that helped make California "The Golden State". **8.5" X 11", 146 ppgs. Retail Price: $12.99**

Mines of San Bernadino and Tulare Counties, California - Originally published in 1917, this important publication on California Mining has not been available for nearly a century. This publication includes important information on the early gold mines of San Bernadino and Tulare County, which were some of the first gold fields mined in California by early Spanish and Mexican miners before the 49ers came on the scene. Included are not only details on early mining methods in the area, production statistics and geological information, but also the location of the early gold mines that helped make California "The Golden State". Also included are details on the mining of other minerals such as copper, iron, lead, zinc, manganese, tungsten, vanadium, asbestos, barite, borax, cement, clay, dolomite, fluospar, gem stones, graphite, marble, salines, petroleum, stronium, talc and others. **8.5" X 11", 200 ppgs. Retail Price: $19.99**

Chromite Mining in The Klamath Mountains of California and Oregon - Unavailable since 1919, this publication was originally compiled by J.S. Diller of the United States Department of Geological Survey and includes details on the chromite mines of this area of Northern California and Southern Oregon. Also included are insights into the mineralization and other characteristics of this important mining region, as well as the location of historic mines. Also included are insights into chromite mining in Eastern Oregon and Montana. **8.5" X 11", 98 ppgs. Retail Price: $9.99**

Mines and Mining in Amador, Calaveras and Tuolumne Counties, California - Unavailable since 1915, this publication was originally compiled by William Tucker and includes details on the mines and mineral resources of this important California mining area. Included are details on the geology, history and important gold mines of the region, as well as insights into other local mineral resources such as asbestos, clay, copper, talc, limestone and others. Also included are insights into the mineralization and other characteristics of this important portion of California's Mother Lode mining region. **8.5" X 11", 198 ppgs. Retail Price: $14.99**

The Cerro Gordo Mining District of Inyo County California - Unavailable since 1963, this publication was originally compiled by the United States Department of Interior. Included are insights into the mineralization and other characteristics of this important mining region of Southern California. Topics include the mining of gold and silver in this important mining district in Inyo County, California, including details on the history, production and locations of the Cerro Gordo Mine, the Morning Star Mine, Estelle Tunnel, Charles Lease Tunnel, Ignacio, Hart, Crosscut Tunnel, Sunset, Upper Newtown, Newtown, Ella, Perseverance, Newsboy, Belmont and other silver and gold mines in the Cerro Gordo Mining District. This volume also includes important insights into the fossil record, geologic formations, faults and other aspects of economic geology in this California mining district. **8.5" X 11", 104 ppgs. Retail Price: $10.99**

Mining in Butte, Lassen, Modoc, Sutter and Tehama Counties of California - Unavailable since 1917, this publication was originally compiled by the United States Department of Interior. Included are insights into the mineralization and other characteristics of this important mining region of California. Topics include the mining of asbestos, chromite, gold, diamonds and manganese in Butte County, the mining of gold and copper in the Hayden Hill and Diamond Mountain mining districts of Lassen County, the mining of coal, salt, copper and gold in the High Grade and Winters mining districts of Modoc County, gold mining in Sutter County and the mining of gold, chromite, manganese and copper in Tehama County. This volume also includes the production records and locations of numerous mines in this important mining region. **8.5" X 11", 114 ppgs. Retail Price: $11.99**

Mines of Trinity County California - Originally published in 1965, this important publication on California Mining has not been available for nearly fifty years. This publication includes important information on mines and mining in Trinity County, California, as well insights into the mineralization and geology of this important mining area in Northern California. Included are extensive details on hardrock and placer gold mines and prospects, including charts showing the locations of these historic mines.. **8.5" X 11", 144 ppgs. Retail Price: $12.99**

Mines of Kern County California - Originally published in 1962, this important publication on California Mining has not been available for nearly fifty years. This publication includes important information on mines and mining in Kern County, California, as well insights into the mineralization and geology of this important mining area in California. Included are extensive details on hardrock and placer gold mines and prospects, including charts showing the locations of these historic mines. **8.5" X 11", 398 ppgs. Retail Price: $24.99**

Mines of Calaveras County California - Originally published in 1962, this important publication on California Mining has not been available for nearly fifty years. This publication includes important information on mines and mining in Calaveras County, California, as well insights into the mineralization and geology of this important mining area in Northern California. Included are extensive details on hardrock and placer gold mines and prospects, including charts showing the locations of these historic mines. **8.5" X 11", 236 ppgs. Retail Price: $19.99**

Lode Gold Mining in Grass Valley California - Unavailable since 1940, this publication was originally compiled by the United States Department of Interior. Included are insights into the gold mineralization and other characteristics of this important mining region of Nevada County, California. This volume also includes important insights into the geologic formations, faults and other aspects of economic geology in this California mining district. Of particular interest are the fine details on many hardrock gold mines in the area, including their locations, histories, development and mineralization. Some of the mines featured include the Gold Hill Mine, Massachusetts Hill, Boundary, Peabody, Golden Center, North Star, Omaha, Lone Jack, Homeward Bound, Hartery, Wisconsin, Allison Ranch, Phoenix, Kate Hayes, W.Y.O.D., Empire, Rich Hill, Daisy Hill, Orleans, Sultana, Centennial, Conlin, Ben Franklin, Crown Point and many others. **8.5" X 11", 148 ppgs. Retail Price: $12.99**

Lode Mining in the Alleghany District of Sierra County California - Unavailable since 1913, this publication was originally compiled by the United States Department of Interior. Included are insights into the mineralization and other characteristics of this important mining region of Sierra County. Included are details on the history, production and locations of numerous hardrock gold mines in this famous California area, including the Tightner Mine, Minnie D., Osceola, Eldorado, Twenty One, Sherman, Kenton, Oriental, Rainbow, Plumbago, Irelan, Gold Canyon, North Fork, Federal, Kate Hardy and others. This volume also includes important insights into the fossil record, geologic formations, faults and other aspects of economic geology in this California mining district. **8.5" X 11", 48 ppgs. Retail Price: $7.99**

Six Months In The Gold Mines During The California Gold Rush - Unavailable since 1850, this important work is a first hand account of one "49'ers" personal experience during the great California Gold Rush, shedding important light on one of the most exciting periods in the history of not only California, but also the world. Compiled from journals written between 1847 and 1849 by E. Gould Buffum, a native of New York, "Six Months In The Gold Mines During The California Gold Rush" offers a rare look into the day to day lives of the people who came to California to work in her gold mines when the state was still a great frontier. **8.5″ X 11″, 290 ppgs. Retail Price: $19.99**

Quartz Mines of the Grass Valley Mining District of California - Unavailable since 1867, this important publication has not been available since those days. This rare publication offers a short dissertation on the early hardrock mines in this important mining district in the California Mother Lode region between the 1850's and 1860's. Also included are hard to find details on the mineralization and locations of these mines, as well as how they were operated in those day. **8.5″ X 11″, 44 ppgs. Retail Price: $8.99**

Alaska Mining Books

Ore Deposits of the Willow Creek Mining District, Alaska - Unavailable since 1954, this hard to find publication includes valuable insights into the Willow Creek Mining District near Hatcher Pass in Alaska. The publication includes insights into the history, geology and locations of the well known mines in the area, including the Gold Cord, Independence, Fern, Mabel, Lonesome, Snowbird, Schroff-O'Neil, High Grade, Marion Twin, Thorpe, Webfoot, Kelly-Willow, Lane, Holland and others. **8.5″ X 11″, 96 ppgs. Retail Price: $9.99**

The Juneau Gold Belt of Alaska - Unavailable since 1906, this hard to find publication includes valuable insights into the gold mines around Juneau, Alaska. The publication includes important details into the history, geology and locations of the well known gold mines and prospects in the area, including those around Windham Bay, Holkham Bay, Port Snettisham, on Grindstone and Rhine Creeks, Gold Creek, Douglas Island, Salmon Creek, Lemon Creek, Nugget Creek, from the Mendenhall River to Berners Bay, McGinnis Creek, Montana Creek, Peterson Creek, Windfall Creek, the Eagle River, Yankee Basin, Yankee Curve, Kowee Creek and elsewhere. Not only are gold placer mines included, but also hardrock gold mines. **8.5″ X 11″, 224 ppgs. Retail Price: $19.99**

Arizona Mining Books

Mines and Mining in Northern Yuma County Arizona - Originally published in 1911, this important publication on Arizona Mining has not been available for over a hundred years. Included are rare insights into the gold, silver, copper and quicksilver mines of Yuma County, Arizona together with hard to find maps and photographs. Some of the mines and mining districts featured include the Planet Copper Mine, Mineral Hill, the Clara Consolidated Mine, Viati Mine, Copper Basin prospect, Bowman Mine, Quartz King, Billy Mack, Carnation, the Wardwell and Osbourne, Valensuella Copper, the Mariquita, Colonial Mine, the French American, the New York-Plomosa, Guadalupe, Lead Camp, Mudersbach Copper Camp, Yellow Bird, the Arizona Northern (Salome Strike), Bonanza (Harqua Hala), Golden Eagle, Hercules, Socorro and others. **8.5″ X 11″, 144 ppgs. Retail Price: $11.99**

The Aravaipa and Stanley Mining Districts of Graham County Arizona - Originally published in 1925, this important publication on Arizona Mining has not been available for nearly ninety years. Included are rare insights into the gold and silver mines of these two important mining districts, together with hard to find maps. **8.5″ X 11″, 140 ppgs. Retail Price: $11.99**

Gold in the Gold Basin and Lost Basin Mining Districts of Mohave County, Arizona - This volume contains rare insights into the geology and gold mineralization of the Gold Basin and Lost Basin Mining Districts of Mohave County, Arizona that will be of benefit to miners and prospectors. Also included is a significant body of information on the gold mines and prospects of this portion of Arizona. This volume is lavishly illustrated with rare photos and mining maps. **8.5″ X 11″, 188 ppgs. Retail Price: $19.99**

Mines of the Jerome and Bradshaw Mountains of Arizona - This important publication on Arizona Mining has not been available for ninety years. This volume contains rare insights into the geology and ore deposits of the Jerome and Bradshaw Mountains of Arizona that will be of benefit to miners and prospectors who work those areas. Included is a significant body of information on the mines and prospects of the Verde, Black Hills, Cherry Creek, Prescott, Walker, Groom Creek, Hassayampa, Bigbug, Turkey Creek, Agua Fria, Black Canyon, Peck, Tiger, Pine Grove, Bradshaw, Tintop, Humbug and Castle Creek Mining Districts. This volume is lavishly illustrated with rare photos and mining maps. **8.5″ X 11″, 218 ppgs. Retail Price: $19.99**

The Ajo Mining District of Pima County Arizona - This important publication on Arizona Mining has not been available for nearly seventy years. This volume contains rare insights into the geology and mineralization of the Ajo Mining District in Pima County, Arizona and in particular the famous New Cornelia Mine. **8.5″ X 11″, 126 ppgs. Retail Price: $11.99**

Mining in the Santa Rita and Patagonia Mountains of Arizona - Originally published in 1915, this important publication on Arizona Mining has not been available for nearly a century. Included are rare insights into hundreds of gold, silver, copper and other mines in this famous Arizona mining area. Details include the locations, geology, history, production and other facts of the mines of this region. 8.5" X 11", 394 ppgs. Retail Price: $24.99

Mining in the Bisbee Quadrangle of Arizona - Originally published in 1906, this important publication on Arizona Mining has not been available for nearly a century. Included are rare insights into hundreds of gold, silver, copper and other mines in this famous Arizona mining area. Details include the locations, geology, history, production and other facts of the mines of this important mining region. 8.5" X 11", 188 ppgs. Retail Price: $14.99

Montana Mining Books

A History of Butte Montana: The World's Greatest Mining Camp - First published in 1900 by H.C. Freeman, this important publication sheds a bright light on one of the most important mining areas in the history of The West. Together with his insights, as well as rare photographs of the periods, Harry Freeman describes Butte and its vicinity from its early beginnings, right up to its flush years when copper flowed from its mines like a river. At the time of publication, Butte, Montana was known worldwide as "The Richest Mining Spot On Earth" and produced not only vast amounts of copper, but also silver, gold and other metals from its mines. Freeman illustrates, with great detail, the most important mines in the vicinity of Butte, providing rare details on their owners, their history and most importantly, how the mines operated and how their treasures were extracted. Of particular interest are the dozens of rare photographs that depict mines such as the famous Anaconda, the Silver Bow, the Smoke House, Moose, Paulin, Buffalo, Little Minah, the Mountain Consolidated, West Greyrock, Cora, the Green Mountain, Diamond, Bell, Parnell, the Neversweat, Nipper, Original and many others. 8.5" X 11", 142 ppgs. Retail Price: $12.99

The Butte Mining District of Montana - This important publication on Montana Mining has not been available for over a century. Included are rare insights into the gold, copper and silver mines of Butte, Montana together with hard to find maps and photographs. Some of the topics include the early history of gold, silver and copper mining in the Butte area, insight into the geology of its mining areas, the local distribution of gold, silver and copper ores, as well their composition and how to identify them. Also included are detailed facts about the mines in the Butte Mining District, including the famous Anaconda Mine, Gagnon, Parrot, Blue Vein, Moscow, Poulin, Stella, Buffalo, Green Mountain, Wake Up Jim, the Diamond-Bell Group, Mountain Consolidated, East Greyrock, West Greyrock, Snowball, Corra, Speculator, Adirondack, Miners Union, the Jessie-Edith May Group, Otisco, Iduna, Colorado, Lizzie, Cambers, Anderson, Hesperus, Preferencia and dozens of others. 8.5" X 11", 298 ppgs. Retail Price: $24.99

Mines of the Helena Mining Region of Montana - This important publication on Montana Mining has not been available for over a century. Included are rare insights into the gold, copper and silver mines of the vicinity of Helena, Montana, including the Marysville Mining District, Elliston Mining District, Rimini Mining District, Helena Mining District, Clancy Mining District, Wickes Mining District, Boulder and Basin Mining Districts and the Elkhorn Mining District. Some of the topics include the early history of gold, silver and copper mining in the Helena area, insight into the geology of its mining areas, the local distribution of gold, silver and copper ores, as well their composition and how to identify them. Also included are detailed facts, history, geology and locations of over one hundred gold, silver and copper mines in the area . 8.5" X 11", 162 ppgs, Retail Price: $14.99

Mines and Geology of the Garnet Range of Montana - This important publication on Montana Mining has not been available for over a century. Included are rare insights into the gold, copper and silver mines of the vicinity of this important mining area of Montana. Some of the topics include the early history of gold, silver and copper mining in the Garnet Mountains, insight into the geology of its mining areas, the local distribution of gold, silver and copper ores, as well their composition and how to identify them. Also included are detailed facts, history, geology and locations of numerous gold, silver and copper mines in the area . 8.5" X 11", 100 ppgs, Retail Price: $11.99

Mines and Geology of the Philipsburg Quadrangle of Montana - This important publication on Montana Mining has not been available for over a century. Included are rare insights into the gold, copper and silver mines of the vicinity of this important mining area of Montana. Some of the topics include the early history of gold, silver and copper mining in the Philipsburg Quadrangle, insight into the geology of its mining areas, the local distribution of gold, silver and copper ores, as well their composition and how to identify them. Also included are detailed facts, history, geology and locations of over one hundred gold, silver and copper mines in the area 8.5" X 11", 290 ppgs, Retail Price: $24.99

Geology of the Marysville Mining District of Montana - Included are rare insights into the mining geology of the Marysville Mining District. Some of the topics include the early history of gold, silver and copper mining in the area, insight into the geology of its mining areas, the local distribution of gold, silver and copper ores, as well their composition and how to identify them. Also included are detailed facts, history, geology and locations of gold, silver and copper mines in the area 8.5" X 11", 198 ppgs, Retail Price: $19.99

The Geology and Mines of Northern Idaho and North Western Montana

See listing under Idaho.

Nevada Mining Books

The Bull Frog Mining District of Nevada - Unavailable since 1910, this publication was originally compiled by the United States Department of Interior. This volume also includes important insights into the geologic formations, faults and other aspects of economic geology in this Nevada mining district. Of particular interest are the fine details on many mines in the area, including their locations, histories, development and mineralization. Some of the mines featured include the National Bank Mine, Providence, Gibraltor, Tramps, Denver, Original Bullfrog, Gold Bar, Mayflower, Homestake-King and other mines and prospects. **8.5" X 11", 152 ppgs, Retail Price: $14.99**

History of the Comstock Lode - Unavailable since 1876, this publication was originally released by John Wiley & Sons. This volume also includes important insights into the famous Comstock Lode of Nevada that represented the first major silver discovery in the United States. During its spectacular run, the Comstock produced over 192 million ounces of silver and 8.2 million ounces of gold. Not only did the Comstock result in one of the largest mining rushes in history and yield immense fortunes for its owners, but it made important contributions to the development of the State of Nevada, as well as neighboring California. Included here are important details on not only the early development and history of the Comstock, but also rare early insight into its mines, ore and its geology.**8.5" X 11", 244 ppgs, Retail Price: $19.99**

Colorado Mining Books

Ores of The Leadville Mining District - Unavailable since 1926, this publication was originally compiled by the United States Department of Interior. This volume also includes important insights into the ores and mineralization of the Leadville Mining District in Colorado. Topics include historic ore prospecting methods, local geology, insights into ore veins and stockworks, the local trend and distribution of ore channels, reverse faults, shattered rock above replacement ore bodies, mineral enrichment in oxidized and sulphide zones and more. **8.5" X 11", 66 ppgs, Retail Price: $8.99**

Mining in Colorado - Unavailable since 1926, this publication was originally compiled by the United States Department of Interior. This volume also includes important insights into the mining history of Colorado from its early beginnings in the 1850's right up to the mid 1920's. Not only is Colorado's gold mining heritage included, but also its silver, copper, lead and zinc mining industry. Each mining area is treated separately, detailing the development of Colorado's mines on a county by county basis. **8.5" X 11", 284 ppgs, Retail Price: $19.99**

Gold Mining in Gilpin County Colorado - Unavailable since 1876, this publication was originally compiled by the Register Steam Printing House of Central City, Colorado. A rare glimpse at the gold mining history and early mines of Gilpin County, Colorado from their first discovery in the 1850's up to the "flush years" of the mid 1870's. Of particular interest is the history of the discovery of gold in Gilpin County and details about the men who made those first strikes. Special focus is given to the early gold mines and first mining districts of the area, many of which are not detailed in other books on Colorado's gold mining history. **8.5" X 11", 156 ppgs, Retail Price: $12.99**

Mining in the Gold Brick Mining District of Colorado - Important insights into the history of the Gold Brick Mining District, as well as its local geography and economic geology. Also included are the histories and locations of historic mines in this important Colorado Mining District, including the Cortland, Carter, Raymond, Gold Links, Sacramento, Bassick, Sandy Hook, Chronicle, Grand Prize, Chloride, Granite Mountain, Lucille, Gray Mountain, Hilltop, Maggie Mitchell, Silver Islet, Revenue, Roosevelt, Carbonate King and others. In addition to hardrock mining, are also included are details on gold placer mining in this portion of Colorado. **8.5" X 11", 140 ppgs, Retail Price: $12.99**

Washington Mining Books

The Republic Mining District of Washington - Unavailable since 1910, this important publication was originally published by the Washington Geologic Survey and has been unavailable for a century. Topics include the geology, rock formations and the formation of ore deposits in this important mining area of Washington State. Also included are hard to find details on the geology, history and locations of dozens of mines in the area. Some of the mines featured include the New Republic Mine, Ben Hur, Morning Glory, the South Republic Mine, Quilp, Surprise, Black Tail, Lone Pine, San Poil, Mountain Lion, Tom Thumb, Elcaliph and many others. **8.5" X 11", 94 ppgs, Retail Price: $10.99**

The Myers Creek and Nighthawk Mining Districts of Washington - Unavailable since 1911, this important publication was originally published by the Washington Geologic Survey and has been unavailable for a century. Topics include the geology, rock formations and the formation of ore deposits in these important mining areas of Washington State. Also included are hard to find details on the geology, history and locations of dozens of mines in the area. Some of the mines featured include the Grant Mine, Monterey, Nip and Tuck, Myers Creek, Number Nine, Neutral, Rainbow, Aztec, Crystal Butte, Apex, Butcher Boy, Molson, Mad River, Olentangy, Delate, Kelsey, Golden Chariot, Okanogan, Ohio, Forty-Ninth Parallel, Nighthawk, Favorite, Little Chopaka, Summit, Number One, California, Peerless, Caaba, Prize Group, Ruby, Mountain Sheep, Golden Zone, Rich Bar, Similkameen, Kimberly, Triune, Hiawatha, Trinity, Hornsilver, Maquae, Bellevue, Bullfrog, Palmer Lake, Ivanhoe, Copper World and many others.
  8.5" X 11", 136 ppgs, **Retail Price: $12.99**

The Blewett Mining District of Washington - Unavailable since 1911, this important publication was originally published by the Washington Geologic Survey and has been unavailable for a century. Topics include the geology, rock formations and the formation of ore deposits in this important mining area of Washington State. Also included are hard to find details on the geology, history and locations of dozens of mines in the area. Some of the mines featured include the Washington Meteor, Alta Vista, Pole Pick, Blinn, North Star, Golden Eagle, Tip Top, Wilder, Golden Guinea, Lucky Queen, Blue Bell, Prospect, Homestake, Lone Rock, Johnson, and others. 8.5" X 11", 134 ppgs, **Retail Price: $12.99**

Silver Mining In Washington - Unavailable since 1955, this important publication was originally published by the Washington Geologic Survey. Featured are the hard to find locations and details pertaining to Washington's silver mines. 8.5" X 11", 180 ppgs, **Retail Price: $15.99**

The Mines of Snohomish County Washington - Unavailable since 1942, this important publication was originally published by the Washington Geologic Survey and has been unavailable for seventy years. Featured are details on a large number of gold, silver, copper, lead and other metallic mineral mines. Included are the locations of each historic mine, along with information on the commodity produced. 8.5" X 11", 98 ppgs, **Retail Price: $10.99**

The Mines of Chelan County Washington - Unavailable since 1943, this important publication was originally published by the Washington Geologic Survey and has been unavailable for seventy years. Featured are details on a large number of gold, silver, copper, lead and other metallic mineral mines. Included are the locations of each historic mine, along with information on the commodity. 8.5" X 11", 88 ppgs, **Retail Price: $9.99**

Metal Mines of Washington - Unavailable since 1921, this important publication was originally published by the Washington Geologic Survey and has been unavailable for nearly ninety years. Widely considered a masterpiece on the Washington Mining Industry, "Metal Mines of Washington" sheds light on the important details of Washington's early mining years. Featured are details on hundreds of gold, silver, copper, lead and other metallic mineral mines. Included are hard to find details on the mineral resources of this state, as well as the locations of historic mines. Lavishly illustrated with maps and historic photos and complete with a glossary to explain any technical terms found in the text, this is one of the most important works on mining in the State of Washington. No prospector or miner should be without it if they are interested in mining in Washington.   8.5" X 11", 396 ppgs, **Retail Price: $24.99**

Gem Stones In Washington - Unavailable since 1949, this important publication was originally published by the Washington Geologic Survey and has been unavailable since first published. Included are details on where to find naturally occurring gem stones in the State of Washington, including quartz crystal, amethyst, smoky quartz, milky quartz, agates, bloodstone, carnelian, chert, flint, jasper, onyx, petrified wood, opal, fire opal, hyalite and others. 8.5" X 11", 54 ppgs, **Retail Price: $8.99**

The Covada Mining District of Washington - Unavailable since 1913, this important publication was originally published by the Washington Geologic Survey and has been unavailable for a century. Topics include the geology, rock formations and the formation of ore deposits in this important mining area of Washington State. Also included are hard to find details on the geology, history and locations of dozens of mines in the area. Some of the mines featured include the Admiral, Advance, Algonkian, Big Bug, Big Chief, Big Joker, Black Hawk, Black Tail, Black Thorn, Captain, Cherokee Strip, Colorado, Dan Patch, Dead Shot, Etta, Good Ore, Greasy Run, Great Scott, Idora, IXL, Jay Bird, Kentucky Bell, King Solomon, Laurel, Laura S, Little Jay, Meteor, Neglected, Northern Light, Old Nell, Plymouth Rock, Polaris, Quandary, Reserve, Shoo Fly, Silver Plume, Three Pines, Vernie, White Rose and dozens of others. 8.5" X 11", 114 ppgs, **Retail Price: $10.99**

The Index Mining District of Washington - Unavailable since 1912, this important publication was originally published by the Washington Geologic Survey and has been unavailable for a century. Topics include the geology, rock formations and the formation of ore deposits in this important mining area of Washington State. Also included are hard to find details on the geology, history and locations of dozens of mines in the area. Some of the mines featured include the Sunset, Non-Pareil, Ethel Consolidated, Kittaning, Merchant, Homestead, Co-operative, Lost Creek, Uncle Sam, Calumet, Florence-Rae, Bitter Creek, Index Peacock, Gunn Peak, Helena, North Star, Buckeye. Copper Bell, Red Cross and others. 8.5" X 11", 114 ppgs, **Retail Price: $11.99**

Mining & Mineral Resources of Stevens County Washington - Unavailable since 1920, this important publication was originally published by the Washington Geologic Survey and has been unavailable for a century. Topics include the geology, rock formations and the formation of ore deposits in these important mining areas of Washington State. Also included are hard to find details on the geology, history and locations of hundreds of mines in the area. 8.5" X 11", 372 ppgs, **Retail Price: $24.99**

The Mines and Geology of the Loomis Quadrangle Okanogan County, Washington - Unavailable since 1972, this important publication was originally published by the Washington Geologic Survey and has been unavailable for a century. Topics include the geology, rock formations and the formation of ore deposits in this important mining area of Washington State. Also included are hard to find details on the geology, history and locations of dozens of gold, copper, silver and other mines in the area. 8.5" X 11", 150 ppgs, **Retail Price: $12.99**

The Conconully Mining District of Okanogan County Washington - Unavailable since 1973, this important publication was originally published by the Washington Geologic Survey and has been unavailable for a century. Topics include the geology, rock formations and the formation of ore deposits in this important mining area of Washington State, which also includes Salmon Creek, Blue Lake and Galena. Also included are hard to find details on the geology, mining history and locations of dozens of mines in the area. Some of the mines include Arlington, Fourth of July, Sonny Boy, First Thought, Last Chance, War Eagle-Peacock, Wheeler, Mohawk, Lone Star, Woo Loo Moo Loo, Keystone, Hughes, Plant-Callahan, Johnny Boy, Leuena, Gubser, John Arthur, Tough Nut, Homestake, Key and many others 8.5" X 11", 68 ppgs, **Retail Price: $8.99**

Wyoming Mining Books

Mining in the Laramie Basin of Wyoming - Unavailable since 1909, this publication was originally compiled by the United States Department of Interior. Also included are insights into the mineralization and other characteristics of this important mining region, especially in regards to coal, limestone, gypsum, bentonite clay, cement, sand, clay and copper. 8.5" X 11", 104 ppgs, **Retail Price: $11.99**

New Mexico Mining Books

The Mogollon Mining District of New Mexico - Unavailable since 1927, this important publication was originally published by the US Department of Interior and has been unavailable for 80 years. Topics include the geology, rock formations and the formation of ore deposits in this important mining area in New Mexico. Of particular focus is information on the history and production of the ore deposits in this area, their form and structure, vein filling, their paragenesis, origins and ore shoots, as well as oxidation and supergene enrichment. Also included are hard to find details, including the descriptions and locations of numerous gold, silver and other types of mines, including the Eureka, Pacific, South Alpine, Great Western, Enterprise, Buffalo, Mountain View, Floride, Gold Dust, Last Chance, Deadwood, Confidence, Maud S., Deep Down, Little Fanney, Trilby, Johnson, Alberta, Comet, Golden Eagle, Cooney, Queen, the Iron Crown, Eberle, Clifton, Andrew Jackson mine, Mascot and others. 8.5" X 11", 144 ppgs, **Retail Price: $12.99**

The Percha Mining District of Kingston New Mexico - Unavailable since 1883, this important publication was originally published by the Kingston Tribune and has been unavailable for over one hundred and thirty five years. Having been written during the earliest years of gold and silver mining in the Percha Mining District, unlike other books on the subject, this work offers the unique perspective of having actually been written while the early mining history of this area was still being made. In fact, the work was written so early in the development of this area that many of the notable mines in the Percha District were less than a few years old and were still being operated by their original discoverers with the same enthusiasm as when they were first located. Included are hard to find details on the very earliest gold and silver mines of this important mining district near Kingston in Sierra County, New Mexico. 8.5" X 11", 68 ppgs, **Retail Price: $9.99**

East Coast Mining Books

The Gold Fields of the Southern Appalachians - Unavailable since 1895, this important publication was originally published by the US Department of Interior and has been unavailable for nearly 120 years. Topics include the geology, rock formations and the formation of ore deposits in this important mining area of the American South. Of particular focus is information on the history and statistics of the ore deposits in this area, their form and structure and veins. Also included are details on the placer gold deposits of the region. The gold fields of the Georgian Belt, Carolinian Belt and the South Mountain Mining District of North Carolina are all treated in descriptive detail. Included are hard to find details, including the descriptions and locations of numerous gold mines in Georgia, North Carolina and elsewhere in the American South. Also included are details on the gold belts of the British Maritime Provinces and the Green Mountains. 8.5" X 11", 104 ppgs, **Retail Price: $9.99**

Gold Rush Tales Series

Millions in Siskiyou County Gold - In this first volume of the "Gold Rush Tales" series, leading mining historian and editor Kerby Jackson, introduces us to the story of how millions of dollars worth of gold was discovered in Siskiyou County during the California Gold Rush. Lavishly illustrated with photos from the 19th Century, this hard to find information was first published in 1897 and sheds important light onto the gold rush era in Siskiyou County, California and the experiences of the men who dug for the gold and actually found it. **8.5" X 11", 82 ppgs, Retail Price: $9.99**

The California Rand in the Days of '49 - In this second volume of the "Gold Rush Tales" series, leading mining historian and editor Kerby Jackson, introduces us to four tales from the California Gold Rush. Lavishly illustrated with photos from the 19th Century, this hard to find information was first published in 1890's and includes the stories of "California's Rand", details about Chinese miners, how one early miner named Baker struck it rich and also the story of Alphonzo Bowers, who invented the first hydraulic gold dredge. **8.5" X 11", 54 ppgs, Retail Price: $9.99**

More Mining Books

Prospecting and Developing A Small Mine - Topics covered include the classification of varying ores, how to take a proper ore sample, the proper reduction of ore samples, alluvial sampling, how to understand geology as it is applied to prospecting and mining, prospecting procedures, methods of ore treatment, the application of drilling and blasting in a small mine and other topics that the small scale miner will find of benefit. **8.5" X 11", 112 ppgs, Retail Price: $11.99**

Timbering For Small Underground Mines - Topics covered include the selection of caps and posts, the treatment of mine timbers, how to install mine timbers, repairing damaged timbers, use of drift supports, headboards, squeeze sets, ore chute construction, mine cribbing, square set timbering methods, the use of steel and concrete sets and other topics that the small underground miner will find of benefit. This volume also includes twenty eight illustrations depicting the proper construction of mine timbering and support systems that greatly enhance the practical usability of the information contained in this small book. **8.5" X 11", 88 ppgs. Retail Price: $10.99**

Timbering and Mining - A classic mining publication on Hard Rock Mining by W.H. Storms. Unavailable since 1909, this rare publication provides an in depth look at American methods of underground mine timbering and mining methods. Topics include the selection and preservation of mine timbers, drifting and drift sets, driving in running ground, structural steel in mine workings, timbering drifts in gravel mines, timbering methods for driving shafts, positioning drill holes in shafts, timbering stations at shafts, drainage, mining large ore bodies by means of open cuts or by the "Glory Hole" system, stoping out ore in flat or low lying veins, use of the "Caving System", stoping in swelling ground, how to stope out large ore bodies, Square Set timbering on the Comstock and its modifications by California miners, the construction of ore chutes, stoping ore bodies by use of the "Block System", how to work dangerous ground, information on the "Delprat System" of stoping without mine timbers, construction and use of headframes and much more. This volume provides a reference into not only practical methods of mining and timbering that may be employed in narrow vein mining by small miners today, but also rare insights into how mines were being worked at the turn of the 19th Century. **8.5" X 11", 288 ppgs. Retail Price: $24.99**

A Study of Ore Deposits For The Practical Miner - Mining historian Kerby Jackson introduces us to a classic mining publication on ore deposits by J.P. Wallace. First published in 1908, it has been unavailable for over a century. Included are important insights into the properties of minerals and their identification, on the occurrence and origin of gold, on gold alloys, insights into gold bearing sulfides such as pyrites and arsenopyrites, on gold bearing vanadium, gold and silver tellurides, lead and mercury tellurides, on silver ores, platinum and iridium, mercury ores, copper ores, lead ores, zinc ores, iron ores, chromium ores, manganese ores, nickel ores, tin ores, tungsten ores and others. Also included are facts regarding rock forming minerals, their composition and occurrences, on igneous, sedimentary, metamorphic and intrusive rocks, as well as how they are geologically disturbed by dikes, flows and faults, as well as the effects of these geologic actions and why they are important to the miner. Written specifically with the common miner and prospector in mind, the book will help to unlock the earth's hidden wealth for you and is written in a simple and concise language that anyone can understand. **8.5" X 11", 366 ppgs. Retail Price: $24.99**

Mine Drainage - Unavailable since 1896, this rare publication provides an in depth look at American methods of underground mine drainage and mining pump systems. This volume provides a reference into not only practical methods of mining drainage that may be employed in narrow vein mining by small miners today, but also rare insights into how mines were being worked at the turn of the 19th Century. **8.5" X 11", 218 ppgs. Retail Price: $24.99**

Fire Assaying Gold, Silver and Lead Ores - Unavailable since 1907, this important publication was originally published by the Mining and Scientific Press and was designed to introduce miners and prospectors of gold, silver and lead to the art of fire assaying. Topics include the fire assaying of ores and products containing gold, silver and lead; the sampling and preparation of ore for an assay; care of the assay office, assay furnaces; crucibles and scorifiers; assay balances; metallic ores; scorification assays; cupelling; parting' crucible assays, the roasting of ores and more. This classic provides a time honored method of assaying put forward in a clear, concise and easy to understand language that will make it a benefit to even beginners. **8.5" X 11", 96 ppgs. Retail Price: $11.99**

Methods of Mine Timbering - Originally published in 1896, this important publication on mining engineering has not been available for nearly a century. Included are rare insights into historical methods of timbering structural support that were used in underground metal mines during the California that still have a practical application for the small scale hardrock miner of today. **8.5" X 11", 94 ppgs. Retail Price: $10.99**

The Enrichment of Copper Sulfide Ores - First published in 1913, it has been unavailable for over a century. Topics include the definition and types of ore enrichment, the oxidation of copper ores, the precipitation of metallic sulfides. Also included are the results of dozens of lab experiments pertaining to the enrichment of sulfide ores that will be of interest to the practical hard rock mine operator in his efforts to release the metallic bounty from his mine's ore. **8.5" X 11", 92 ppgs. Retail Price: $9.99**

A Study of Magmatic Sulfide Ores - Unavailable since 1914, this rare publication provides an in depth look at magmatic sulfide ores. Some of the topics included are the definition and classification of magmatic ores, descriptions of some magmatic sulfide ore deposits known at the time of publication including copper and nickel bearing pyrrohitic ore bodies, chalcopyrite-bornite deposits, pyritic deposits, magnetite-ileminite deposits, chromite deposits and magmatic iron ore deposits. Also included are details on how to recognize these types of ore deposits while prospecting for valuable hardrock minerals. **8.5" X 11", 138 ppgs. Retail Price: $11.99**

The Cyanide Process of Gold Recovery - Unavailable since 1894 and released under the name "The Cyanide Process: Its Practical Application and Economical Results", this rare publication provides an in depth look at the early use of cyanide leaching for gold recovery from hardrock mine ores. This volume provides a reference into the early development and use of cyanide leaching to recover gold. **8.5" X 11", 162 ppgs. Retail Price: $14.99**

California Gold Milling Practices - Unavailable since 1895 and released under the name "California Gold Practices", this rare publication provides an in depth look at early methods of milling used to reduce gold ores in California during the late 19th century. This volume provides a reference into the early development and use of milling equipment during the earliest years of the California Gold Rush up to the age of the Industrial Revolution. Much of the information still applies today and will be of use to small scale miners engaging in hardrock mining. **8.5" X 11", 104 ppgs. Retail Price: $10.99**

www.ingramcontent.com/pod-product-compliance
Lightning Source LLC
Chambersburg PA
CBHW080806180526
45168CB00006B/2343